The Coyote
Defiant Songdog of the West

THE AUTHOR, FRANÇOIS LEYDET, is known for his writing on nature and conservation subjects. His previous books, *The Last Redwoods*, and *Time And The River Flowing: Grand Canyon*, have been lauded for their eloquence and knowledge. His byline has also appeared on articles in National Geographic, American West, Vogue and Holiday magazines, among others.

Born in France, raised in Boston and educated at Harvard, Leydet moved West in the 1950's, soon became a U.S. citizen and an ardent advocate of preserving America's wilderness tradition. Kindled by early stints as shrimpboat crewman in the Gulf of Mexico, and ranch hand in eastern Oregon, Leydet's love of our outdoor heritage has led him to interrupt his career as author and newspaperman with frequent expeditions as river guide/boatman on the Green and Colorado rivers.

The Coyote
Defiant Songdog of the West

By François Leydet

Illustrations by Lewis E. Jones

Chronicle Books / San Francisco

Library of Congress Cataloging in Publication Data

Leydet, François, 1927–
    The coyote.

        1. Coyotes—Behavior.   2. Coyotes—Control.
3. Predation (Biology)   I. Title.
QL737.C22L49   599'.74442   76-55396
ISBN: 0-87701-086-2

Chronicle Books
870 Market Street
San Francisco, Ca. 94102

Permissions to Quote

To ROZ,
my beloved wife,
and to
JOHN CORNELIUS MORRIS,
my patient friend.
But for their caring and believing
this book would not have been born

Contents

Acknowledgements

To me, the acknowledgments section is the most difficult part of a book to write. Perhaps other authors, too, find it to be so.

For one thing, I have but very little space in which to thank a great many people, and the recognition I give them is bound to appear perfunctory. I can only say to all who have helped me in the preparation of my book that these brief credits are not the measure of my gratitude.

For another thing, I have the nagging feeling that, no matter how careful I am, I will omit mention of one or more individuals who richly deserve recognition. To them what can I say—other than to plead human fallibility and to extend my sincere apologies!

My primary debt is to the Weatherhead Foundation of New York, which in 1970 awarded me a very generous research grant that enabled me to devote a major part of my time and energy to this book. It was the Foundation's support, above all, that made the book possible. One of the Weatherhead Foundation's principal areas of concern is the ecology of the Southwest. It has supported projects that, in the words of Foundation president Richard W. Weatherhead, "seek to foster a better understanding of the nature of the desert and mountain environment in which man must live with the flora and fauna and life-giving air and water." I shall be ever grateful to the Weatherhead Foundation for deciding that my book, in which I proposed to study the relationship of two predators, man and the coyote, to this environment and to each other, was germane to the Foundation's sphere of interest. I am further grateful to the Foundation for its patience in repeatedly granting me extensions for the completion of my task.

I also wish to express my thanks to the Marin Conservation League, and to its then president, Harold Gregg, for playing a vital role in the implementation of the grant.

To some extent, this book is indirectly related to research I did in the 1960's on the general subject of predators and predator control. For support of that early work I am beholden to the Sierra

Club, National Audubon Society, Defenders of Wildlife, Belvedere Scientific Fund, Mrs. Arthur Hays Sulzberger, Nathaniel and Margaret Owings, Kenneth L. Lindsey, Alfred E. Heller, and Walther H. Buchen. Dr. A. Starker Leopold, professor of zoology at the University of California in Berkeley, gave generously of his time and advice. To all of them I give my belated thanks.

As will become clear in the reading of this book, a great many individuals of all kinds of expertise took time and pains to enlighten me about the coyote. Many are mentioned by name in the book. Each contributed in his or her own way and has earned my lasting gratitude; to all of them I extend my thanks. I hesitate to single out any one of these individuals lest I seem to disparage the help I received from the others, but I do want to specifically thank Vern Cunningham of the U.S. Fish and Wildlife Service, state supervisor for Animal Damage Control in New Mexico, not only for keeping his office door open to me, but also for arranging for me to observe the trapping and aerial hunting scenes I describe in the book.

Among those who helped but are not mentioned in the text, I wish to thank my friends William B. Franklin of Tucson, Mrs. Harris Carrigan of Pass Christian, Mississippi, Thomas Fleming of Mill Valley, California, and Peter Phillips, of Belvedere, California; conservationist E. D. Marshall of Ventura, California; retired trappers Harold Dobyns of Pendleton, Oregon, and Lester Barton of Deer Lodge, Montana; environmentalist Pauline Derry of Roy, Utah; Dr. John A. Kadlec and Dr. Frederick H. Wagner of Utah State University in Logan, biologists and former members of the Cain Committee; outdoor writer Bill Davidson of Alamosa, Colorado; Dr. Paul B. Sears, ecologist, of Taos, N.M.; Colonel Henry Zeller of Santa Fe, environmentalist; Dr. Florence Ellis of Santa Fe, anthropologist; Walt Snyder, Louis Berghofer, and Gerald Gates of the New Mexico Department of Game and Fish; Harley Shaw, research biologist for the Arizona Game and Fish Department; Frank Lamb, Max Cummings, and biologist Louis C. Cox of the U.S. Fish and Wildlife Service; Ed Olson, animal damage control specialist, Window Rock, Arizona; Dr. Karl W. Luckert, anthropologist, and graduate biology student Jim Witham, at the University of Northern Arizona in Flagstaff; Dr. Hermann Bleibtreu, director of the Museum of Northern Arizona; Drs. Ed Carpenter, Russell Gum, and Norman Smith at the University of Arizona in Tucson, and doctoral candidates Dennis Danner and Al Fisher; Carlos Nagel of the Arizona-Sonora Desert Museum, Tucson; Edward C. Rodriguez, Jr., superintendent, Organ Pipe Cactus National Monument, Arizona; John Riffey of the National Park Service, Tuweep, Arizona; M. P. "Pete" Espil of Tolleson, Arizona, president of the Arizona Woolgrowers Association; Oliver "Sato" Lee, of Reserve, N.M., vice president of the New Mexico Cattlegrowers Association; Gary Bogue, curator of the Alexander Lindsay Junior Museum, Walnut Creek, California; and Albert E. LaRose, of the Marin County, California, Department of Agriculture.

Nearly two hundred men and women around the nation and in Canada took the trouble to write to me in response to a letter to the editor which I placed in a large number of newspapers and magazines, and in which I asked for personal experiences with, and opinions about, predatory animals. Some wrote at great length, and their letters were full of interesting details—a compilation of them would make a fascinating book in itself. It pained me that space allowed me to quote only a tiny fraction of these letters, and that lack of time prevented me from even acknowledging receipt of most of them. Such correspondents as Robert D. Hanson, Sacramento, California, Norman E. White, Quincy, California, Joe Frasier, Woodrow, Colorado, Lewis P. Gunnell, Pocatello, Idaho, Norman Tague, Boise, Idaho, Dean Parisian, Hancock, Minnesota, John T. Talbott, Cedar Hill, Tennessee, Charles A. Clements, Houston, Texas, Thomas D. Smith, Burnet, Texas, Maurice L. Nelson, Windsor, Vermont, Eve Smith, South Pender Island, British Columbia, Robert W. Page, Winnipeg, Manitoba—may think me ungrateful although I am anything but that! Their letters and the many like them provided me with invaluable background information, and the writers have my very sincere thanks.

The text contains a number of quotations from works by other authors, who said certain things better than I could have said them myself. To them, and to their publishers, I am grateful for permission to quote. I particularly want to thank Dr. Samuel B. Linhart, research biologist for the U.S. Fish and Wildlife Service, for permission to reproduce, as Appendix A, the entire text of his paper, *Progress in Coyote Depredations Research*.

I am greatly indebted to Robert D. Roughton of the Fish and Wildlife Service for his perceptive and painstaking annotated criticism of my manuscript; to Dr. A. Starker Leopold who also read it and made valuable suggestions; to Harry O. King, Jr., research associate, the Wheelright Museum, Santa Fe, and Richard Randall of Rock Springs, Wyoming, retired government trapper and field representative for Defenders of Wildlife, for reviewing the sections relevant to their special areas of expertise; and to my friends Dr. Robert L. Martensen of San Francisco and the Rev. Henry Parsons King of Manchester, Massachusetts, for their literary criticisms of the finished work.

To my editor, Phelps Dewey, go my thanks for the patience, confidence, and skill with which he assisted in the long gestation and birthing of this work.

On a more personal note, I wish to express my profound gratitude to my dear friend and father-in-law, Charles R. Carney of Manchester, Massachusetts, for his generous response to my eleventh hour appeal for help: without his aid, which allowed me to spend another six months on the task, this book very literally would not have been written. I also thank my step-children, Walter E. Robb, IV, Rachel Robb, Holly Harris, and Mary-Peck Harris, for their unswerving support during the preparation of the book and their good-humored understanding of my moods and distances.

Finally, there are two persons whose support was most crucial of all, and to whom I have dedicated this book: John Cornelius Morris, M.D., who patiently guided me through a Kierkegaardian "dark night of the soul," and my beloved wife, Roz, who made the journey worthwhile.

F. L.
Kentfield, California
October, 1976

¡Cantad, Amigos!

CLEAN AND UNTRODDEN, the beach fringed a cove of the strait that separates Tiburón Island from the Sonoran mainland. The big, uninhabited island, which bulked starkly across the horizon to the west and southwest, sheltered this stretch of coast from the winds blowing across the Sea of Cortés. The water's surface was as smooth as silk, the air soft and balmy even though the month was only February. An osprey left its perch in a cardón—Mexico's larger counterpart to Arizona's giant saguaro—and dove for fish in the cove. My German shepherds, two-year-old Hamlet and four-month-old Ophelia, dug holes in the sand, delighting in the freedom of this beach after the long drive from Tucson. I, too, liked the setting and decided that we would camp here for the next two or three days.

We were unlikely to have any human visitors. The last twenty miles of coastal road had been a narrow, sandy track that followed the curves of the shore and wound its way between cardones and ocotillos—the latter flaunting flamelike flower clusters at the tips of their tall, spiny stems. The nearest outpost of civilization was Bahía Kino, a small resort and fishing port some thirty miles south by east. The nearest human settlement that I knew of, about twenty miles down the coast, was a Seri Indian village which I had driven by that afternoon. A group of its inhabitants had stopped my car and offered for my inspection and purchase small animal figures carved from ironwood, nearly as heavy as bronze, almost as dark and polished as the sculptors' skins. For a few pesos I had bought a sea turtle as a present for the young lady I was soon to marry; now, six years later, it resides in our living room.

At sunset, as I grilled a steak over a fire of dead mesquite, I watched flights of pelicans cruising low above the dark water. Before sinking behind Tiburón Island, the sun gilded the sere brown hills of Sonora, to my left. The gentle lapping and susurration of wavelets against the sand made the only sound. Until well after dark I sat by the embers of the campfire, savoring the peace and beauty of the deserted scene. The dogs dozed quietly nearby, tired from the drive and their wild romping on the beach.

Finally I kicked sand over the coals, gave my friends a hug, and climbed into my sleeping bag. Even then I lay awake, staring at the myriad stars glittering with miraculous brightness in the unsullied sky. Occasionally a meteor would streak among them in a fiery arc; it had no competition from blinking airplane lights, as shooting stars so often do in Arizona or California. In this wild, lonely spot, so untouched by any of the manifestations of man, I felt totally serene, content, and safe.

When, out of the black, Hamlet sat up and rumbled a low, muted growl, I mocked his alarm: "What's the matter, boy? There's nobody here!" He ignored me and sat motionless, ears pricked forward, looking intently at the line of low trees and brush that fringed the beach. I looked too, but I could see and hear no movement of any kind, no sign of anything suspicious. Again Hamlet growled, this time more authoritatively. Ophelia whimpered and snuggled close to me. Then, from the darkness just back of the beach, there exploded a single, staccato, high-pitched bark. Other single barks accelerated into multiples, like bursts from a machine gun. The cacophony resolved into musical tones, escalating into long, wavering soprano notes. First one voice, then another, then several in chorus carried or harmonized the melody. Hamlet was clearly worried. His hackles were up and he never moved from my side. I expected him to join in the howling, just as he usually responded to a fire siren, but he confined himself to low rumbles. Ophelia was plainly terrified, and to soothe her I took her into my sleeping bag.

After my own initial startle I began to feel elated. What could be more perfect in this setting than a wild chorus of coyotes? I switched on my flashlight for a few seconds, shone it in the direction of the disembodied voices, and counted eight pairs of eyes reflecting its beam. I shut it off, and began to bark and howl back at our visitors, an action which seemed to spur them on to even more exuberant polyphony. "¡Cantad, amigos!", I called out, the directive to sing welling up from a subconscious memory. The coyotes continued their song for a few moments, then, almost as suddenly as it had started, the chorus stilled. In a couple of minutes I saw Hamlet relax, telling me that our callers had left.

What my subconscious had remembered, I recognized after a while, was a scene at the beginning of J. Frank Dobie's classic, *The Voice of the Coyote*. The setting for that scene was "a small ranch down in the Brush Country of Texas." A steer had been butchered

LEWIS JONES

that afternoon. Some of the meat had been taken to the rancher's and ranch hands' households; some had been hung in the open for sun and wind to cure into jerky; the offal had been dragged out of the corral into the mesquites, where the coyotes would feast on it in the night. "Well, we'll have singing tonight," the rancher said after sundown, as he sat down with his family to a supper of fried steak, hot biscuits, and gravy. Dobie explains:

> The speaker was my father. The particular beef-killing day I am remembering was forty-odd years ago. Fresh meat and singing went together in the same place forty or fifty years before that; they still go together. After a group of vaqueros— as Mexican cowboys in the border country are called—fill up on fresh beef in the evening, they are going to sing, and if they have been freshly brought together for a cow hunt, they are going to sing more. When coyotes smell fresh meat after dark, they also are going to sing.
>
> "Well, we'll have singing tonight." We did not have long to wait. High-wailing and long-drawn-out, the notes of native *versos* came over the night air . . . [and] seemed to go up to the stars. And as they reached their highest pitch, a chorus of coyote voices joined them. When, at the end of the first ballad, the human voices dimmed into silence, the coyote voices grew higher; then all but one howler ceased. We heard a laugh, and one lusty vaquero yelled out, "*¡Cantad, amigos!*" —"Sing, friends!" The friends responded with renewed gusto.

"*¡Cantad, amigos!*" The words themselves seem to sing, not only because Spanish is a musical language, but also because these words bespeak a joyful attitude of harmony with the natural world, of camaraderie with fellow creatures. The rancher, the vaqueros, the coyotes had shared the flesh of the slaughtered steer, and now they shared their contentment in a burst of song that echoed back and forth in the starry Texas night. Dobie continues:

> For a time the antiphonies challenged and cheered each other, now converging, now alternately lapsing. The vaquero singing, on high notes especially, could hardly be distinguished from the coyote singing. Mexican vaqueros are Indian in blood, inheritance and instinct. Anybody who has listened long and intimately to the cries of coyotes seeming to express remembrance of something lost before time began must recognize those identical cries in the chants of Plains

Indians and the homemade songs of Mexicans. English-speakers living with the coyote seldom refer to his voicing as "singing"; to them it is "yelping," "howling," "barking"; but to the vaquero people it is nearly always *cantando*.

Dobie's father, the rancher, must have shared something of his hands' attitude towards their "amigos," for it was surely he who gave the order to drag the steer's paunch and entrails into the mesquites where the coyotes could dine on them. He evidently did not view them as alien trespassers; they, too, were natives of this land, with a right of place and a right to partake of its bounty. And apparently his gesture of letting the little songdogs have their go at what was left of the steer was not unusual in the early days of this century, as two old-timers indicated in letters they wrote to me.

One letter came from Frank Jemmett, 79, of Melba, Idaho. He wrote:

I have worked with sheep and cattle and have owned both all my life. . . .We used to run as high as two and three thousand ewes in a band . . . them days there was all ways something for a coyote to eat . . . there was a dead cow or a dead horse . . . a sheep here and there for to eat on. . . .Today people has become so greedy and money hungry that they gather up all the bones and now sell them. . . .

Another letter, from 85-year-old Emmett Douglass of Bozeman, Montana, confirms this:

I started in the Sheep in June 1914, and have had from 1,500 to 6,000 head . . . in my early days . . . from 1909 when we lost a cow or horse on the Ranch or on the Range they were left there for what ever wanted to eat them. Today when we loose an animal we call the rendering plant and they come and pick them up.

But note the difference in tone between the two letters, a difference that will become the central focus of this book. Jemmett writes:

I never owned a cow that would let a coyote kill her calf. . . Idaho was a large sheep and cattle country in the early days. The sheep men always built up their herds and prospered they never went broke from coyote like they claim to day and I think I would be truthful in saying that there was a 100 coyote then to where there is one to day. Not only that we had bear and wolves mountain lions etc. But we went ahead and built up. We had herders that took care of their herds. . . . Besides them days there was all ways something for a coyote to eat. . . That is something they dont have to day you can go for a week and never see a carcus any more. Coyotes were put here for the balance of nature to clean up things like that, hold down the squirrels and rabbits etc. . . .

Poision should not be used. They done that before it got so bad you could not keep a good dog. It was so bad that we had to use muzzles on our dogs to keep them from getting poisioned. I'll tell you Mr. if you was ever caught out on the range and sawn the best dog you ever owned get a bait and die and you could not help him and was left alone with a band of sheep, I'll tell you how I felt if I could of run accross the man that droped it I would of gladley and willingly shot him through the middle and left him to rot there. . . .

Douglass, on the other hand, while agreeing that "there isn't the feed for the cyote that there was years back, there is not any rabits to speak of and we don't have the field mice," says bluntly that:

. . . there is only one solution to cyote predation on sheep and only one and that is 1080 [a virulent chain-acting poison injected into baits]. *If the sportsman want anything to shoot at they had better cooperate with the ranchers and get rid of the cyote and the Fox. . . .There is verry few people that knows what they are talking about when they are talking cyote, Montana could and will be a good good sheep country if they just get rid of the cyotes and never before. . . .*

Or contrast, if you will, the tone in these two letters: Pat Wheat wrote to me from Boulder, Colorado:

Everything that lives preys on something else; and the most wanton, ruthless, mindlessly destructive predator of all is MAN. So what are you talking about?

Then, Mrs. Tom Carey, a young rancher's wife from Boulder, Montana, began her letter as follows:

Many years ago, Frank Carey, who was one of the earliest settlers in the Boulder Valley (settled in the 1860s), made the remark, "If I saw one coyote and I knew he was the last coyote on earth, I'd shoot him anyway." Today . . . we wish it had been the last coyote and that he had shot and killed it.

What makes one man cry out happily, *"¡Cantad, amigos!"* while another passes sentence, "Just get rid of the coyotes"? This book, in its attempt to answer this central question, deals not with one animal but two—*Canis latrans*, the coyote, and *Homo sapiens*, man—and studies the interaction of these two most successful large predatory species in the West.

In the course of gathering material for this book, during the early months of 1976, I "talked coyote" with dozens of individuals of different backgrounds, attitudes, and expertise, but all having some tie to or interest in the coyote: sheepmen and cowmen, sportsmen and environmentalists, state and federal wildlife researchers and "government trappers," university professors, and Indians. I saw dozens of coyotes. I saw them play, I saw them hunt, I saw them mate, I saw them die. I heard men praise the coyote, I heard them revile him, I watched them study him, I watched them kill him.

And in all of my conversations I made it a point to ask not only, "What do you *know* about the coyote," but also, "How do you *feel* about him?" The answers often were unexpectedly revealing. When one rancher growled that "these worthless welfare coyotes we've got today are too lazy to hunt their natural prey," he was giving me an insight into more than just his attitude towards *Canis latrans*. I often found that what a man told me about his feelings towards the coyote revealed much about his whole perception of man's place and role on this Earth. Without necessarily using the terms, some spoke of dominion, others of stewardship.

In Berkeley, Dr. A. Starker Leopold, University of California biologist and conservationist, and an author of the famous 1964 "Leopold Report" on predator control, cautioned me, "What you really need to bring out in your book is how *little* is known about the coyote." Certainly the more I learned about the coyote the more I realized how little I myself knew. Thus this book makes no pretense of being a definitive work, but, unless you happen to be a coyote expert, you may here learn something about this enigmatic animal. You may learn something about your fellowman. And you might even, as I did while writing it, learn something about yourself.

2

Neither Cruelty
Nor Compassion

THE LUNCH HAD been delicious. Sipping my coffee, I sat in drowsy contentment in a comfortable armchair, one ear listening to the animated conversation between my wife and our hostess, the other tuned to the strains of a Mozart piano concerto wafting gently from the stereo. A warm spring sun poured in from the garden and brightened the soft tones of a Navajo rug. Suddenly, from the corner of my eye I caught a flash of movement outside: a blur of blue, red, white, and black. I stood up, intrigued, and walked to the window. It took me a moment, unprepared as I was, to realize that a few feet outside the glass, in the civilized precincts of this California patio and at the foot of an uncaring live oak, a savage drama was being played out. A scrub jay was hammering away with its powerful, pointed beak at a cowering rufous-sided towhee. Peck, peck, peck, peck, again and again the blue bird stabbed at the towhee's head and back, while the slightly smaller red-black-and-white bird made feeble and futile attempts to flutter away.

As soon as I grasped the nature of the scene, I found myself wrenched by two opposing impulses. Part of me wanted to rush out and rescue the towhee. Another part of me cautioned, "Who am I to play God?" After a moment's hesitation the Good Samaritan instinct won out, and I hurried out the front door. The jay saw me at once and flew a short distance away, jeering or protesting in a series of harsh cries. I could see that the towhee was badly hurt. It no longer made any effort to fly; its head and back glistened wetly, whether from blood or the jay's spittle I could not tell. When its head slowly sank forward, lifted up weakly, then sank again, I decided the bird was too far gone to be helped. I backed into the doorway, still watching.

In a moment the jay was back. It pecked at the towhee two or three more times, then tightened its beak around the little neck and carted its victim away in flight. I shut the front door and returned to my seat, feeling somewhat shaken. I had not expected to look out on this grisly scene when I stood up to investigate that

flash of color outside the window. Still less would I have antici-
pated, although I knew that jays are omnivorous and do take fledg-
lings of other birds, that I would see one prey on an adult bird
nearly its own size. Shock, even some anger, were the feelings that
stayed with me.

I had just witnessed violence, presumably fear and pain, and, I
assumed, death. My sympathies lay with the victim, yet, would it
not have been more natural for me to identify with the jay? Certainly I had more in common with him than with the towhee: both
the jay and I were still alive. Also, had I not just lunched on
chicken that, having been served to me plucked, gutted, dismembered, broiled, and masked with a delectable sauce, still represented the carcass of a bird that must have met death not long
before in a frenzy of squawking, flapping terror? Had I not been
the ultimate cause of that chicken's death just as the jay had been
the proximate cause of the towhee's? Why should I be shocked at
an instance of predation when I myself was a predator, a carnivore
who, in nearly a half century's consumption of meat, had occasioned the deaths of countless beefs, calves, lambs, pigs, chickens,
turkeys, geese, ducks, and fish?

I suppose I am as ambivalent, even confused, in my attitude
towards predation as most of my fellows. It is difficult to overcome
the feeling that predation is somehow evil, sinful. Like so many
others, I was brought up on *The Three Little Pigs* and *Little Red
Riding Hood*. I, too, cheered for Stickly-Prickly and Slow-and-
Solid against Painted Jaguar when Mother read me Kipling's *Just
So Stories*. I, too, grew up thinking of wolves and other large
predators as cruel, vicious, wanton killers. I, too, pictured deer
and rabbits and songbirds as innocent, timorous animals haunted
by the constant dread of being eaten alive.

It took me many years and much study to clear my mind of these
childhood fantasies and prejudices. Many of us never do outgrow
them. One who did was naturalist Sally Carrighar, author of *Wild
Heritage*. In the chapter titled "How Red the Tooth and Claw?"
she recalls that "during a youthful and idealistic period I wished pas-
sionately that all animals might be vegetarians." But she notes that:

> One must spend many months in a wilderness to see even a
> few attacks. They take place without warning and usually are
> over so quickly that dependable observations are hard to
> make. Finally, after watching enough of them, the neutral
> quality of the pursuits becomes evident. Perhaps it is not

really important whether one dies between murderous, vicious jaws or jaws that are merely hungry; yet to me it was a moment of great relief when I realized that man's expression "blood-lust" means no more than the literal ache of an empty stomach. . . .

Carrighar recalls vividly the first fatal encounter she watched.

    . . . A fox was the hunter. Tawny-coated, he was enjoying himself in the limber green grass of a meadow. Rather idly he seemed to be looking for food, since he sniffed the ground here and there and sniffed the breeze; but his mood was light, and his movements suggested that he was taking pleasure in his own grace. He heard the click of a tree squirrel's claws and raced towards the oak, barking, although he must have known he could not reach the squirrel. A brown towhee was scratching under a bush, jumping forward and back on parallel feet.

The fox bounded towards it, not so fast but that the bird had a chance to fly up and away. As the fox returned to the meadow, a dragonfly, conspicuous with the sunlight stitched into its wings, skimmed over his head. He leapt for it but not eagerly—that too was only a bit of play. Very suddenly, then, the fox became still, listening, with his eyes on the ground and his motionless head slightly cocked. With brilliant speed he springs into the air and comes down with all his feet clustered into a grasping unit. He has caught his prey, a young ground squirrel, perhaps. I could not see what it was, but it was more than a mouse, for it went down in several bites. Meanwhile the fox, as he chewed, seemed strangely casual, almost uninterested. Here was a revelation: that fox didn't hate his victim any more than I hate the lamb when I watch a butcher wrap up some chops.

Golden Shadows, Flying Hooves, George Schaller's book on African lions, similarly describes the predator's affect on seizing prey:

> Merely to watch the constantly changing facial expressions of lions when they interact or respond to their surroundings is a fascinating way to spend the hours. The face of a lion is a marvelously subtle yet clear conveyor of emotions. . . .They express themselves so unambiguously, their features and sounds mirror their minds so precisely, that only the most insensitive of persons could grossly misinterpret them. . . . When a cub chases another cub, slaps it, pulls it down, and bites its throat, are the actions derived from hunting prey or from fighting between lions, circumstances that involve quite different emotions? These forms of behavior can be readily distinguished among adults by facial expressions alone: prey is captured unemotionally, silently, and with bland features, whereas bared teeth and ferocious snarls accompany fights.

On two occasions I witnessed scenes similar to the one Carrighar describes, except that my hunters were coyotes. One occurred in a broad snowfield at the foot of the San Francisco Peaks, near Flagstaff, Arizona. The other took place on the National Elk Refuge in Jackson Hole, Wyoming. Both times a coyote was hunting mice under the snow. I particularly remember the latter one because he was an especially handsome animal—those Jackson

Hole coyotes seem much larger, better-fleshed, and heavier-coated than the coyotes of the Southwest deserts—and I spent a good forty-five minutes watching him through my field glasses from a distance of a hundred yards, while I shivered amid light snow flurries. He was working a low mound covered with a thick blanket of snow. His total concentration, his patience, his muscular control, his grace were remarkable; so was his impassivity: he was going about the job of earning a living in a totally matter-of-fact, untheatrical way. For minutes on end he would stand absolutely motionless, looking down at the mound, ears pricked forward, head slightly cocked to one side. Then, like Carrighar's fox, he would suddenly spring straight up and come down with all four feet together in a marvelously quick, elastic, seemingly effortless pounce. He would bury his face in the snow at his feet, and when he straightened up his eyes and nose were three black specks in a mask of white. Several times his pounces were futile, but three times I could tell by his snapping jaws that he had collected a meal. That he could hear the tiny squeaks and rustling of the mice through a foot or more of snow was impressive testimony to the acuity of his hearing.

I noticed another interesting thing that day at the National Elk Refuge. An average of 7,500 elk winter every year in its broad meadows. To observe the elk from close-by, the public is allowed to take sleigh rides through the midst of the herds. Earlier that afternoon I climbed aboard a heavy sled drawn by two massive Belgian horses, the teamster shook the reins, and we took off in the direction of the elk. These were not massed in one contiguous herd, but rather were scattered, in groups of a few hundred, over the wide flat ground of the refuge. We headed straight for one such group. Only when we were less than a hundred yards from them did the cows move to the side, some walking, some trotting. The lordly bulls, with their extravagant racks of antlers, hardly moved at all. Some lay calmly in the snow, chewing their cuds, as we passed only a few yards away. "What would happen if I stepped off the sled?," I asked the teamster. "Oh, then there'd be a stampede," he answered. "As long as you're aboard they don't seem to know you're human. But as soon as you walk away from it you then become a hunter."

Real hunters were about that day: in the course of the ride I counted fourteen coyotes trotting across the refuge. The elk paid them no apparent heed—they evidently knew that the coyotes were not hunting *them*. Indeed, the coyotes do not bother the elk on the refuge, although they might dispatch one that was in some

way disabled, and they do perform the service of quickly cleaning up the remains of any elk that dies. But some prey species apparently can distinguish between a predator on the hunt and one occupied with other matters, and they react accordingly. My elk were unawed by the coyotes, and the mere sight of a wolf will not stampede a herd of caribou. Zebra will continue to graze with an eye on a pride of lions and will take flight only when the lions show by their movements or by crossing an invisible boundary line that they are on the prowl.

"There is neither cruelty nor compassion," Schaller writes, "in a lion's quest for food . . . most hunts [are] struggles of life and death at their most elemental. It is a time when each animal uses to the utmost those attributes with which evolution has endowed it. . . ." Then he discusses fear:

> The life of a prey animal always hangs in the balance, for it never knows if it has made a serious error until the final rush. Yet prey is not constantly alert; it does not anticipate momentary death. I wonder if these animals have fear as we know it. Fear is a subjective sensation, an anticipation of future feelings based on past events, but few animals have ever suffered the agony of a mauling and lived. No, the prey flees primarily because tradition passed down from its kind dictates that it do so when a lion is close and its heritage commands it to retreat when it sees and perhaps smells certain things. Prey leads a mundane existence of feeding, resting, traveling, courting, fighting. Evasive action is taken when needed. If a member of the herd dies, the others watch the event, then calmly continue with their routine as if conversant with the inevitability of death.

Our reactions to predation seem to vary considerably, depending on who the predator is, and who the prey. The lower we rate an animal on evolution's scale, the less affected we seem by its death. Big fish eats little fish—we shrug. One spring, when I was camped by a dry wash in a remote section of Organ Pipe National Monument in Arizona, I watched a huge wasp, a "tarantula hawk," attack and subdue a tarantula, then drag it back towards its burrow. I watched quite unperturbed, although I realized that by our lights this was a particularly nasty type of predation: the tarantula was not dead, only paralyzed, and its death would come later, slowly, as the wasp's larvae fed on its vital fluids.

I have made many river trips through the Grand Canyon, rowing a wooden dory, and I sometimes have been asked if I ever see any predators down there. I tend to say very few and to tell of the young bobcat I saw sitting atop a rock on the shore near Three Springs Canyon. He watched us drift by as unconcernedly as a house cat in a window might watch pedestrians on the sidewalk outside. I tell of another I saw on the right bank below Gneiss Canyon Rapid, this one a big tom who retreated without panic up a steep draw as we rowed by. I mention sighting tracks of ringtail cats on streamside beaches, of finding an occasional fox or coyote scat, of the final night of one trip when, camped on a beach at the head of Lake Mead, and bathed by the light of a full moon, we were treated to a rollicking serenade by what sounded like a horde of coyotes ringing our camp. In all probability, given the ventriloquial abilities of the animals, they numbered three or four at the most.

A couple of bobcats, a few pawprints, one coyote chorus are not much in the way of predator sightings for ten river runs totalling 2,800 miles. And so, when asked if I have seen many predators in the depths of the Inner Gorge, I reply, "Very few." This is, of course, nonsense. On every day of every river trip I have seen predators by the score and have witnessed acts of predation. Those lovely violet green swallows, swooping and looping after small insects, or skimming the surface of quiet stretches of river, briefly raising their wings into a V, and taking a quick sip of water—what are they but predators on the prowl? In the balmy evening hours, after the sun has slid over the rim of the canyon, the pipistrelles issue from their rocky roosts to perform their aerobatics in the dimming light. Their flight seems erratic, but it is far from that: their every sudden swerve and dip leads them to a moth which their sonar has detected in flight. These little bats, too, are consummate predators.

And what of the myriad lizards—the collared, desert spiny, whiptail, and all their cousins, which race across the sands or do push-ups on sunbaked rocks? Yes, of course, they are predators, too, all but the vegetarian chuckwalla. I remember one especially well. I had beached my dory after running a very rough rapid and was lying on the hot sand drying out, when I saw a red ant crawling up my leg. Tired from the rowing and stupefied by the 110-degree temperature, I made no move to brush it off. But I was not the only one who had noticed the ant. Suddenly, out of the shade of a tamarisk, a desert spiny lizard raced across the sand, darted up my leg, snapped up the ant, and scurried off. "Thank you," I murmured. Remembering the favor, I was a little sorry, on a later trip, to watch another spiny lizard disappear down the throat of a king snake. But this, I knew, was the way it was supposed to be, and so I did not interfere.

On a word-association test, the average person, given the word "predator," will most likely reply "lion," "tiger," or "wolf." Yet a predator is *any* animal that kills another for food, and the ladybug is as lethal a killer as the leopard. But though we may remain impassive at the sight or thought of an insect, a reptile, a fish being devoured by a predator, we tend to empathize with the victim when it belongs, as we do, to a warm-blooded species, when it is a bird or a mammal. It is much easier for us to clothe ourselves vicariously with fur or feathers than with scales or chitin; we instinctively recognize a kinship with our warm-blooded cousins, no matter how distantly related we may be, forgetting our pre-mammalian ancestry that traces back through reptiles and amphibians to slimy creatures of the deep. We are much closer, phylogenetically, to sea squirts than to angels, and that worm in your lawn, desperately resisting the tug of the robin's beak, is as surely your cousin as its bright-eyed capturer. "The robin's getting his breakfast," you smile approvingly, but let your neighbor's cat try to make a breakfast of the robin and your reaction will be quite different!

Our revulsion at predation becomes tinged with indignation when a "lower" animal presumes to make a dinner of its betters. A pike surging from the bottom of a pond to grab a swimming duckling, a sidewinder ingesting a kangaroo rat, a crocodile engulfing a thrashing, screaming human, there seems to be something obscene, certainly anarchical, about such happenings, as if the consumers had forgotten their station in life. I, too, have these feelings, and given as I sometimes am to macabre speculations, I long since have decided that in the unlikely event that I should end my days in a predator's jaws, I should much rather be recycled into tiger flesh than shark meat. In his magnificent book *All the Strange Hours* Loren Eiseley tells of a friend of his youth, Chou Li Han, of whom it was said that he had been eaten by a tiger in the Tibetan highlands. "I thought of Han," Eiseley writes, "toiling relentlessly on and on through the Himalayan solitudes. A great yellow-eyed beast had come slowly to meet him in the snow. . . ." To me, there could be much more ignoble ways to die. But a *shark?*

Indeed, we are taxonomic snobs. We sit self-enthroned at the pinnacle of the pyramid of life: "What a piece of work is a man!" Hamlet cried. "How noble in reason! how infinite in faculty! in form, in moving, how express and admirable! in action how like an angel! in apprehension how like a god! the beauty of the world! the paragon of animals!" (He did add, however, "And yet, to me, what is this quintessence of dust? Man delights me not, no, nor woman neither. . . .") And from our lofty pedestal we survey animal kind and assign higher or lower ranks to all other species. In this arbitrary, anthropocentric hierarchy, the upper levels are reserved to those species which we recognize to be most like ourselves. We have other classifications, too, equally anthropocentric or anthropomorphic: useful or useless, good or bad, peaceful or cruel, graceful or ungainly, beautiful or disgusting. A wolf cub is "cute," its parents are "vicious killers." "How asinine!" one might say, and thus slander the asses!

But Nature is no snob. She is egalitarian. She knows no higher animals, no lower, no good, no bad, no useful, no harmful. Everything that lives is the equal of all others, everything that breathes has its place, has its role. There is a time to eat and a time to be eaten, and individuals are necessarily sacrificed in the process. The life force is endlessly recycled from species to species, as Hamlet himself recognized:

> HAMLET: Your worm is your only emperor for diet: we fat all creatures else to fat us, and we fat ourselves for maggots: your fat king and your lean beggar is but variable service; two dishes, but one table: that's the end.
> KING: Alas, alas!
> HAMLET: A man may fish with the worm that hath eat of a king, and eat of the fish that hath fed on that worm.

Nature seems only to recognize the successful and the unsuccessful. On the former she bestows her favors; the latter, in time, she jettisons. Who are we to say that, except in our own eyes, we are a higher form of life than the scorpion who, talk about a success story, has survived without essential change since Silurian times? In his book *Grand Canyon* Joseph Wood Krutch wrote about the trilobites, very primitive crustaceans whose fossils are found in the Grand Canyon cliffs:

> Trilobites flourished for perhaps 200,000,000 years. They were the dominant animal during a considerable part of that era, and no other animal has ever been for so long a time the kingpin of creation. . . . They lasted far longer than the

dinosaurs were to last—to say nothing of man who, so far, has only a very brief history and may, for all we know, have only a very brief future. Trilobites had much better reason to suppose that the earth belonged to them than we have to assume that it belongs to us or that it was created primarily *for* us.

The trilobite was not the first predator, and man is unlikely to be the last. Predation is as integral a part of life as the double helix. Almost every animal is prey to several others and, for that matter, there are carnivorous plants. But even if one accepts the inevitability of predation and the rightfulness of the predator's role in the web of life, he is apt to be shocked by individual acts of predation, as I was shocked by the jay's attacking the towhee. And our sympathies often are highly selective. The human deer hunter condemns the coyote for eating a fawn. The sheepman—sincerely outraged by the suffering of the lamb torn by the coyote's fangs— can be quite insensitive to the suffering of the coyote whose mangled leg is locked in a trap. The preservationist can state glibly, as one did to me, that "sheep just love to die," and be moved to tears by pictures of poisoned coyotes.

When there is clear physical superiority of predator to prey, the victim's suffering can be minimal. Those mice I saw taken by my Jackson Hole coyote must have died in mid-squeak, never knowing what bit them. But the larger the prey, compared to the predator, or the more inexperienced the predator, the less is death likely to be quick, "clean," and "merciful." Most abhorrent to us is the sight of a predator feeding on its still-breathing prey. But as Robert D. Roughton, research biologist for the U.S. Fish and Wildlife Service, wrote to me,

. . . *If one step is to accept, and educate others to accept, that predators must kill to live, then it seems but another step to realize that they do so dispassionately. For all we know a predator may revel in the chase (some seem to), but once the animal is down it becomes a food item in the same sense that a tender twig is to a deer. There may be a lesson for us if we can observe a coyote pulling the entrails out of its living meal and overcome our repugnance. All of animal life is pragmatic. A predator needs to kill before eating only if the meal is going to move around inconveniently or, for those with some means of physical defense, cause injury to the predator in the course of its meal.*

I spent a day in Rock Springs, Wyoming, talking with Richard Randall, field representative for Defenders of Wildlife, a conservation organization based in Washington, D.C. For many years, before disillusionment and air-crash injuries caused him to change jobs, Dick had been a trapper for the Fish and Wildlife Service. He retired from the service in 1973 as the Rock Springs district's acting supervisor, in charge of the federal predator control program.

"Predation," he said, "is something that if the average person, city person, witnessed they'd probably throw up! Because it isn't something where you run out and slit the animal's throat and he bleeds to death and then you eat. You eat him regardless of whether he's dead or not, you get that piece of meat down, when you're starving you start eating, and oh, boy, you get someone to witness that, some innocent from the city, and oh! We attach too many human values to animal behavior. Right away we could say, This is sadistic! But there's no such thing as sadism in the animal world."

Dick also recalled an aerial hunt during which, "We were tracking a coyote in the snow. The first thing we noticed that was odd was a big hole in the snow, and then the coyote tracks took out from there wobbling a little bit and kind of erratic, and just ahead of us we saw something black, came down over it, and there was an eagle sitting on a coyote. The coyote was still alive. We could figure out exactly what had happened—that the eagle had hit the coyote once and knocked him down, and hit him a second time and probably broken his back, because the coyote could get up on one paw and turn his head back but he was unable to get up on his feet. And the eagle had started eating on the coyote's rear end, and the coyote was watching the whole proceedings. Well, that's predation!"

I myself was witness to an equally stomach-churning scene, and the fact that I saw it on film, rather than in the field, did not make it any more palatable. I was visiting the Albuquerque headquarters of the U.S. Fish and Wildlife Service's Region 2, which embraces Arizona, New Mexico, Oklahoma, and Texas. Terry E. Anderson, assistant regional supervisor for Animal Damage Control, showed me a screening of a twelve-minute movie prepared for the service's Division of Wildlife Research. Filmed in Utah, it showed attacks by two different coyotes on two tiny lambs.

Terry had warned me that the film was gory, but I had asked to see it, and I describe it now because I believe it is essential to see and portray predation in a realistic, objective way, knowing and acknowledging its mixture of horror and beauty. The first killing in the film was fairly quick; it appeared to me in fact that the lamb suffered as much by being trampled by its mother in her ineffectual efforts to shield it from the coyote as it did from the coyote's teeth. The second killing was ghastly. The coyote, a young bitch, seemed hesitant at first to move towards her prey, but after much sniffing around and a stop or two to urinate she finally made a lunge for the tiny lamb. The ewe charged; the coyote danced away. Feint and countercharge alternated for perhaps a minute. Then the coyote managed to slip by the ewe's guard, seized the lamb by a leg, and ran with it a few yards. The ewe charged again, the coyote dropped the lamb and backed away a few feet.

Blood flecked the lamb's wool, and although the film was silent you could see that it was bleating in terror. Twice more the coyote succeeded in dodging past the ewe and dragged the lamb some distance through the snow. After the third time the ewe just gave up. She rejoined another sheep who had remained passive in the background, and now the coyote was free to kill the lamb unmolested. A more experienced coyote might have dispatched the death blow swiftly with a bite in the throat or through the top of the head. But although the lamb was down, kicking feebly now and then, this coyote did not seem to know how to finish it off. Instead she proceeded to eat the prey, first chewing off its ears. "Kill it! Kill it! For God's sake kill it!" I implored the coyote under my breath as I sat with fists clenched, fighting the urge to shut my eyes. More and more blood stained the snow around the lamb, and soon its face was lost to a viscous red mask, as the coyote gnawed at the flesh on its head. Though the lamb had stopped kicking, I could see its little flank still heaving as the reel spun to its end.

"Well, what did you think of it?", Terry Anderson asked, after we had walked back to his office. I could not repress a shudder. "I thought it was God-awful," I replied truthfully, if understatedly. "I had to *force* myself to watch it to the end. But hell, if you and I had only our teeth with which to procure our meat, we'd botch the job even worse than the coyote did. And though I hated watching it, I could no more hate that coyote for wanting to kill that poor lamb than I will hate *myself* the next time I roast a leg of lamb."

"That's right," Terry said, "you can't *hate* an animal for doing what nature has meant it to do. The coyote, you, and I, we're all in this predator business together."

That evening, back at my motel, I sought to balance the ugliness of what I had seen by drowning myself in the splendor of the *Messiah*, played on my cassette recorder. Great music, to me, has some of the properties that occasionally, and all too briefly, can be found in ethyl alcohol. It can soothe you, ease your pain, free you from your mortal coil, give wings to your thoughts, kindle love in your heart, transfigure your soul.

But on this particular evening I found the *Messiah*, for all its glory, less soothing than usual. I had wanted, for what was left of that day, to get that lamb out of my mind. But the constant allusion to sheep in the Biblical passages which Handel had used made this impossible. "He shall feed His flock like a shepherd, and He shall gather the lambs with His arm, and carry them in His bosom, and gently lead them that are with young," sings the alto. "Are we like sheep have gone astray?" asks the choir. "Worthy is the Lamb that was slain," proclaims the final chorus. Yes, we *are* like sheep, in more ways than one, I thought. Had I walked alone in the wrong section of Albuquerque that night, I might have been as fair game to human predators as that lamb was to the coyote, although better able to defend myself. And worthy indeed was that lamb that was slain, I reflected, worthy of a fairer chance at life than being put into a pen with a hungry coyote. But then, as my friend Sister Mary Neill, O.P., likes to say, "Who ever said life was fair?"

I kept worrying the question, just how much *did* that lamb suffer? How much actual physical pain did it feel before it died? This speculation continued to haunt me until, just recently, I asked a number of physician and veterinarian friends of mine. Their consensus, after listening to my description of the filmed attack, was that very early in its mauling the lamb would have gone into shock and thereafter felt no pain.

The most unequivocal in his diagnosis was Robert L. Martensen, a young doctor with extensive experience in emergency-room medicine at a large San Francisco hospital. Almost daily he has treated patients brought in after suffering severe traumas: accidents, failed suicides, shootings and knifings. "That first bite would have been painful," he said. "After that there would have been no pain. No doubt that lamb *was* bleating in terror. And terror, like great anger or excitement or other powerful emotions, is an analgesic."

Many a patient had come in to the emergency room, he told me, bleeding and in shock from grievous wounds—a leg cut off, shot in the chest, stabbed in the abdomen. "I've been cut," one might say. "How much pain are you having?" "I don't feel much." Just recently one man was brought in who obviously was desperately ill. While in the emergency room he suffered a ruptured aortal aneurism. "This should have been exquisitely painful," my friend said. "But all he said, before they rushed him to the operating

room where they miraculously saved his life, was, 'I'm filling up.'
No pain. He was in shock."

I asked my veterinarian friend, Dr. Henry P. Boyd, if the lamb
would have gone into psychological shock before physiological
shock set in. "You can't separate the two," he answered. "Psychology, neurology, physiology, it's all part of one process."

Incomprehension and disbelief are the usual reactions of severely injured humans, and presumably of animals, too. And with those
go the lowered blood pressure, peripheral circulatory inadequacy,
and the resultant analgesia that characterize physiological shock—
nature's extreme unction, her way of enabling her creatures to
bear the unbearable.

I have been saying, in effect, predation was in the beginning, is
now, and ever shall be; it is a fact of life, and like it or not we have
no choice but to accept it. But predation has its more positive,
beneficial, even aesthetic aspects. The restraining effect of predators on prey populations, and vice-versa, are essential forces in
the ever-shifting balance of nature. But here I wish to emphasize
that much that we find beautiful, stirring, surprising, and intriguing in the animal world is directly related to the ubiquitous
influence of predation.

Consider, for example, the fascinating phenomena of mimesis
and mimicry. Biologists studying the myriad strands that mesh to
form the web of life long have been awed by its intricate and pervasive pattern of deception.

At almost every level along the food chain, organisms attempt to
remain unobserved by their predators or prey, or, if that is impossible, to appear to be something other than what they are. Inver-
tebrates, particularly insects, have developed camouflage and
masquerade into a fine art. They have adapted their colors,
patterns, and shapes to copy with amazing precision particular
elements of their environments. In a Costa Rica rain forest, one
chrysomelid beetle looks exactly like a sun-spangled raindrop on a
leaf. Certain caterpillars, such as that of an Arizona giant swallowtail butterfly, look precisely like bird droppings. Some tropical
preying mantises convincingly ape dead leaves or live flowers.
Membracid bugs resemble thorns, berries, leaf buds, and bracts.

Cryptic coloration operates as a first line of defense. When a
predator threatens to breach this line, when it sees through the
insect's camouflage and prepares to attack, there are quite a few
animals (certain mantises, bugs, beetles, butterflies, and moths)

equipped with a secondary defense system: bright spots that closely resemble vertebrate eyes. In experimental studies, birds have been observed to jump back violently when butterflies suddenly opened their wings and flashed their eye spots—as if the birds had come face to face with their own predators, such as cats or owls.

In contrast to those animals that try to remain inconspicuous, many animals are so gaudily colored that they positively invite detection. Brightly painted butterflies; shiny, metallic-hued beetles; rainbow-banded snakes; brilliantly plumed birds; and gaily colored reef fishes, are clearly visible to potential predators, many of whom—amphibians, lizards, birds, monkeys, fish—have good-to-excellent color vision. In some species, bright coloration is related to courtship, but in many it serves a different purpose: it advertises to would-be predators that the gaudily decorated animal is ill tasting or venomous. This is known as warning, or aposematic, coloration.

An inexperienced toad, stung by a hornet, will remember the insect's black-and-yellow banding, associate that color pattern with the painful experience, and avoid insects so marked in the future. Predators have no instinct to avoid these colors; each generation of young learns by unhappy experience.

The effectiveness of aposematic coloration has led nature further in evolving one of her most amazing deceptive tricks: true mimicry, as distinguished from crypsis or mimesis as the foregoing adaptations are termed. The first clear exposition of the mimicry theory was set forth in 1862 by a young Englishman, H. W. Bates. While wandering extensively through the Amazonian jungles, Bates had noticed that certain brightly colored, Heliconiid butterflies, which are unpalatable to birds, were very closely imitated by certain Pieridae, which birds find very much to their taste. And these Pieridae differed greatly in appearance from their near relatives. Bates reasoned that predators, having learned to respect the warning patterns of unpalatable or poisonous insects, would avoid similarly patterned but palatable ones as well.

In the cast of a Batesian mimicry ring there are, therefore, at least three actors: There is the animal that is avoided for good reason, and that is termed the model; there is the imitating animal with the false warning pattern, the mimic; and there is the animal, usually a predator, that recognizes and is interested in the warning pattern or signal shared by model and mimic, and is known as the signal receiver or mimefactor. Genuine aposematic coloring benefits both model and signal receiver: the wasp does not want to be eaten, the bird does not want to be stung. But with the false warning pattern the advantage is one-way: In being fooled by a hover fly's wasplike appearance the bird misses a meal, but the fly escapes with its life.

I cannot go into all the astounding intricacies of mimicry—Müllerian mimicry, dimorphism, polymorphism, auto-mimicry, aggressive mimicry. Perhaps more than any other phenomenon, mimicry proves the variability of species and the impact of natural selection on their evolution. But I have dwelt on the subject because mimicry, to me, represents the most spectacular manifestation in nature of the response of animals to the force of predation.

Among mammals, however, true mimicry is unknown, with the possible exception of mimetic pairs of tree shrews and squirrels described in 1916 by R. Shelford in *A Naturalist in Borneo.* Mammals and birds have as effective, if cruder, means of escaping detection, generally in the category of cryptic coloration. Many arctic animals, for instance, have white coats, at least during the winter months. To polar bears, arctic foxes, snowy owls, snowshoe hares, ptarmigans, this camouflage is of benefit both to the hunters (bears, foxes, owls) and to the hunted (hares, ptarmigans). Similarly, many animals of the American West, inhabiting generally barren, rocky, or sandy terrains, are dun or tan colored and blend almost perfectly into their surroundings: A bighorn sheep against a talus, a coyote on an alluvial fan, a jackrabbit in a sagebrush flat, are practically imperceptible except at close range or unless they move. Even so showy an animal, when seen in a zoo, as a tiger or jaguar is well-camouflaged in its natural habitat of tall dry grasses or sun-dappled jungle forest floor.

Cryptic coloration, of course, is only hide-deep. But predation has had impacts that are far more profound than that. You can measure the power of the wolf's jaws by the angle of the deer's hip. You can gauge the lightness of the coyote's tread by the length of the jackrabbit's ears. What use has the deer for its steel-spring legs, its alacritous web of nerves, the protective embrace of the herd, but that it, like its forebears, must ever be ready to flee for its life? What need had the wolf to evolve its keen senses, its bone-crunching jaws, its indefatigable lope, its high order of intelligence, its fidelity to the pack, but for the biological imperative to outwit, outrun, and subdue prey larger than itself?

Nor is the human hunter, with his snowmobiles and scope-sighted rifles, ever likely to supplant the evolutionary influence of the wolf. Deer and wolf have co-evolved down through the patient aeons, each testing and refining the other's capabilities. Evolution advances on tiptoe, by cautious, tiny steps. No quantum leaps for her, such as might enable the deer to outspeed the bullet! When the human hunter sings praises of the deer, lauds its beauty and alertness, its fleetness and grace, he ought never to forget to pay homage to the wolf. For the wolf is father to the deer, as the deer is mother to the wolf, and in this quasi-incestuous relationship the taking of deer by wolf is an act of consummation, as much as of consumption.

I do not contend that predation has been the *only* evolutionary force. But massive environmental changes, the advance and retreat of the seas, the uplifting and leveling of mountain chains, the progression and recession of the ice, seem to have engendered great extinctions rather than great creations—except where the extinction of species may have left niches vacant for others to fill.

But in the day-to-day, millennium-to-millennium onward crawl of life, I think it safe to say that predation has been the prime mover of animal evolution. And we, you and I, have been as moulded by its pressures as that scatomorphic caterpillar of the Arizona giant swallowtail. Whether, as Robert Ardrey argues, we are descended from carnivorous apes, or whether our proto-human ancestors were peaceful and we only later acquired, along with erect posture, a taste for red meat, does not matter here. What matters is that it was in response to predatory urges, our own or others', that we developed our nimble hands, our weapons-making skills, our agile and fertile brains. Our imperatives were no different from those of any other species: to eat, to procreate, and to avoid being eaten. It was only our *method* of coping with these imperatives that was unique, at least in degree: the enormous development of our cerebral functions, which in time led to language, and to the arts, and to a vision of God.

I recognize our uniqueness, but I reject our alleged separateness. It is lonely up there atop the pyramid of life; a cold wind blows in from the intergalactic void, and I gladly step down to a lower, less-exposed level and cuddle up to the warmth of my brother animals. When I speak of my cousin the toad or my brother the coyote, I mean just that. And I am ever saddened and dismayed to hear so many of my fellow-humans denying this commonality. This hubris, I fear, may be our undoing in the end.

The Voice of the Desert

IN THE BEGINNING there is the darkness, and the warmth, and the comfort. Snuggled with his siblings against his mother's flank, the coyote pup soon recovers from the trauma of birth. To his unawakened senses, the dry world of the den has much in common with the wet world of the womb. What little light reaches the inner chamber, dug fifteen feet back from the mouth of the den, does not register on the newborn pup: like his cousins the wolf and the dog, or his more distant relatives of the cat family, he was brought into the world blind, his eyes tightly shut. His mother's body heat, the soft forms of his litter mates squirming against him, and his own dense, fuzzy coat insulate him from the cool April air that penetrates the den at night from the high desert of northern Arizona.

One sensation only, unfamiliar and gnawing, disturbs his sleepy comfort. It comes from somewhere within him, a pang of emptiness, and it nags him until he feels impelled to bestir himself. Groping instinctively along his mother's belly, his muzzle finds a small, hard nipple. He begins to suckle, and as his forepaws knead the swollen teat he feels a rich, warm liquid fill his mouth and throat. At first he sputters and chokes, but his pharynx quickly learns its function, he swallows, and sucks, and swallows, and in short time the emptiness is gone. He relinquishes the nipple, tired by his exertions; he feels the gentle caress of his mother's tongue, stroking his face and body; and with a little sigh he sinks back into sleep.

Life is a simple matter, these first days, for the pup and his brother and two sisters. Most of the time they sleep, bundled together in a formless heap of soft, gray fur. No dreams disturb their slumber—their conscious lives are still too amorphous to be reflected by their unconscious. They awake only to nurse, move their limbs, and be groomed by the mother coyote. Even so, this is a period of discovery for the pup and his fellows. His senses of hearing and smell begin to relay messages to his brain, a brain endowed with a capacity for learning and intelligent thought probably surpassed only in the primates and the cetaceans: "The coyote is the smartest person next to God," declares an old Mexican saying.

The pup's first feeding has awakened his powers of smell and taste, to remain closely allied in him throughout his life. Whatever he can scent, he will also want to explore by taste. "Call Thou Nothing Unclean," J. Frank Dobie titled his chapter on the coyote's food habits. Before long, the pup's inquisitive mouth is tasting not only his mother's milk but his brother's ear, a sister's tail, his own forepaw.

His siblings' hungry mewlings, his own petulant whimpers as he fights to shoulder past them to his favorite nipple, become familiar sounds to him. Occasional muffled noises from the outside world reach the inner sanctum of the den, but they are devoid of associations and meaningless to the pup. One such noise, however, he comes to recognize though not yet to identify, for it affects the routine of the den. Two or three times a day a sharp, staccato sound causes him to twitch his not-yet-erect ears. The pup cannot know that this is the bark of his father, returning with food for the famished mother coyote. He does notice, however, that his mother always answers the call with a special whine; he feels her briefly nose him, then brush past and be gone. He notices also that on her return her breath usually exhales a pungent odor, an odor which in time he will know as that of flesh and of blood, the scent of life itself, the scent of one life taken that another might go on.

Gradually the mother's absences become more prolonged, until after the fifth day she enters the den only to nurse her brood. Her visits are not always so prompt or frequent as the pup might wish. The first time he awakes with hunger in his belly and finds no ready fountain of milk, he whimpers in puzzlement, anxiety, and frustration. The source of nourishment, he discovers, behaves independently of his desires; it is something more than a mere extension of himself. Similarly, although he can command his own limbs to move with ever-growing strength and coordination, his will has no such instant power over the motions of his litter mates, who regularly jostle with him for position when feeding time comes. He becomes aware of "I" and "they," as later he will distinguish between "I, coyote," and "it, rabbit."

(Indeed, a coyote pup's awareness of its environs develops amazingly fast. Harking back to his coyote-killing days, during which he had to destroy many a den-ful of pups, now Defenders of Wildlife representative Dick Randall said to me: "They're a *coyote* when they're *born*. They're not a little soft, cuddly thing. I've dug into dens in the sand, and got back to the little pocket where the

pups were, and they'd not have their eyes open, and several times I've had one wobbly little pup sit up in the middle of that whole bunch and actually wrinkle his nose and snarl! He didn't know who the hell I was, but he *knew* something was wrong. And *that* makes you back off and think, when you know you're going to bump him in the head in a minute. . . . *I* don't know; some people are soft-hearted and some aren't, I guess.")

The tenth day of life adds a wonderful new dimension to the pup's growing consciousness. He turns again and again towards the exit tunnel, that shaft his mother uses in her comings and goings and through which only cool air and dampened sounds have touched his privacy. But today he feels an expectancy he cannot understand. His black button nose, his mobile little ears, twitch inquiringly, but they give him no clue as to what there is about the tunnel that commands his attention. Little by little, his sealed eyelids relax their bond, and a sensation, ever so faint at first, of lambency, of release and expansion flows through his being. This sensation grows until midafternoon when his lids finally part, and suddenly his world of darkness becomes a world of light.

It is dim light, to be sure, in the back of the den, and in the beginning the pup is baffled by the shadows and shapes and movement that surround him. He gazes uncertainly about, and blinks at the apparition he sees coming at him through the tunnel: a pointed, black-tipped snout followed by two piercing slanted eyes that seem to transfix him with their stare. He cringes as the snout reaches him and sniffs him up and down, then suddenly he relaxes. For if the mother coyote is startling to his eyes she is happily familiar to his nose. There is the odor of milk and meat and sun-warmed fur, the scent that means food and comfort. His growing brain makes the connection between sight and smell. Equipped now with all his senses, the pup learns to coordinate them as he assesses the world around him: whatever he hears he must try to see, what he sees he must smell, what he smells he must taste.

In the next days the pup comes to know every square foot of the rock, dirt, and sand that form the floor, ceiling, and walls of the inner den. He learns that periods of light alternate regularly with periods of darkness. He studies the details of his mother's anatomy, from the gleaming curve of her canines to the thick brush of her tail. He sees, without reflecting on it, that there is something odd about her right foreleg: it has lost its foot, and ends in a stub on which black scar tissue has replaced the normal covering of short fur.

To account for this, we must flash back nine months to the previous summer, when the bitch was still an inexperienced yearling. Her home grounds at the time were fifteen miles away, in a valley that dipped gently northward from the ponderosa-clad shoulder of a volcanic mountain. There a band of sheep, a remnant of once-numerous flocks, competed with a scattered herd of whiteface cattle for the abused and ever-sparser desert grasses, brush, and cacti. The coyote herself had ample food—jackrabbits and cottontails, kangaroo rats and meadow mice, ground squirrels, insects, and occasional lizards. In season, she could feast on the sweet, juicy tunas of the prickly pear, or the plentiful sweet-and-sour berries of the juniper trees.

The cattle and sheep would supplement her diet even if she never attacked a single head of stock. Already, without killing, the she-coyote had profited from the presence of the cattle. At calving time, the cows' afterbirths had provided her with several nourishing meals, and she had relished the calves' droppings, rich in undigested milk. Every now and then throughout the year, she might run across a windfall in the guise of a dead steer or sheep.

Early on a July morning, as dawn pearled the sky over the eastern rim of the valley, the coyote trotted across the flat towards the opposite slope where she would doze away the day in the shade of a juniper, a hundred feet up from the baked valley floor. Depending on his point of view, a man who saw her at this moment would have described her as mangy, scrawny, and sneaky—or simply short-coated, slight, and weary. She was, in fact, a small, doglike figure in thin summer pelage with head and tail carried low, indeed weary and more than a little hungry, for this particular night's hunting had netted her one mouse. But as she neared the rising ground at the western edge of the valley, the dawn breeze wafted a scent to her ever-questing nose that brought her up short, suddenly curious and intent. She faced left into the breeze, studied its message, then started up the valley whence the scent seemed to come. A half mile's trot brought her to the bank of an erosion gully, and looking down into it she saw the object of her search: a dead ewe.

Despite her hunger, she did not go directly to the carcass. In typical coyote fashion she first carefully studied the lay of the land. Her ears on the alert and her nostrils working, she scanned the lightening desert with her sharp yellow eyes, watched the placid grazing of a bunch of sheep a few hundred yards away, marked the takeoff of a meadowlark from across the gully. She looked again at the dead ewe—its hide already had been torn open by vultures and ravens the day before. She then crossed the gully some distance from the sheep and resumed her inspection from the opposite bank. Finally satisfied that the situation held no threat, she jumped down into the gully, walked daintily up to the carcass, sniffed it over thoroughly from the vacant eyes to the dirt-matted tail, then bit into the rump, tearing off big chunks of strong tasting flesh which she gulped down whole without pausing to chew.

In less than five minutes she had eaten her fill—four pounds of meat, a fifth of her own body weight. Sated, belly bulging, panting slightly from her bolted meal, she backed away from the remains, licking the taste off her chops. She scrambled somewhat heavily up the side of the gully, jogged away a few paces then threw herself to the ground, rolling luxuriously onto her back. She thrashed about like a beheaded snake, kicked at the air like an upended beetle, flopped to one side and the other, pulling herself along the ground on her flank, and rubbed her face in the dust to wash off the gore. Then she jumped up, shook herself, and stood for a moment gazing at the sun, which had just topped the far rim of the valley and caused every sagebrush plant to cast a juniper-length shadow.

It was gallant to be a coyote just then, young as the day was young, unfettered as the breeze, with the strength that had been stored in the tissues of the dead sheep now coursing through her body. A mindless need welled up inside her to proclaim to the sun and to the answering hills that she was alive and found life good. Lifting her muzzle towards the luminous sky, she burst into a peal of frenzied, high-pitched barks, an urgent cadenza that ached to resolve into the tonic chord and inevitably did: eyes closed, ears laid back, mouth rounded into an O, she poured forth her feelings in a long, tremulous howl that floated out over the valley, a wild, exultant yet melancholy song that momentarily spooked the grazing sheep and awakened in the dog, at the sheep camp three miles away, dim echoes of a past long forgotten by his race.

Her spirit appeased, the coyote retraced her steps down the valley, stopping once at a stock tank for a long, thirsty drink, then turned west and climbed to her hillside retreat. She lay down in the open, rested her chin on her forepaws, and let the warmth of the still low-hanging sun relax her tired muscles and drug her into sleep. But as the sun climbed higher its heat grew more intense, and well before mid-morning the coyote got up, stretched langorously, gazed briefly at an object crawling along the valley floor and trailing a plume of dust (the object meant man, but was too distant to worry her), and flopped down in the shade of a juniper to resume her slumber.

The pickup lurched off the dirt road that transected the valley, bounced through the sagebrush, and came to a halt by the side of the erosion gully. The man jumped out, studied the sky again, and concluded that whatever was attracting the vultures must be close by. He called the dog, who leaped eagerly to the ground from the bed of the pickup, and set off along the bank of the gully towards a place where he had seen a vulture apparently about to land until scared off by the truck.

After a short walk the man and dog reached the spot on the bank where the coyote had first stood, and six large birds with naked red heads and necks took to the air with a loud flapping of wings. Seeing the sheep they had been feeding on, the man cursed and hurled a hard clump of earth after them. He commanded the dog to stay put, and jumped down into the gully to inspect the torn carcass. Almost at once, he spied small dog-like tracks among the vultures' footprints in the sandy bed of the gully. He noted the deep wound in the sheep's rump, associated it with the canine tracks, and without further investigation rendered his diagnosis of the ewe's demise. "Coyote!" he spat, and the ancient predator-hatred of his kind welled up hot and bitter within him.

There was no room in his world for coyote or bobcat, no room for the golden eagle cruising the sky on eight-foot wings, no room for the hawk or the night-haunting owl. He found no beauty in their beings, no usefulness in their lives, no call for respect in their insubmission to man. Such animals were mere killers, marked with the brand of Cain, and poison, trap, and bullet were their due. At every opportunity he cursed Richard Nixon for the executive order which, two years before, had banned the use of poisons on public lands. Now he fetched his gear from the pickup, donned cotton gloves and, kneeling on a drop cloth, disposed three blind traps around the carcass, covering each carefully with sifted earth and sand. A fourth trap he concealed at the spot where prints showed the coyote had landed when it leaped down from the bank. The chances were, he figured, that it would approach that way again. Satisfied with his work he walked back to the pickup and drove away, intending to return in two or three days to check results.

The coyote did not go to the carcass that night, neither did she hunt. She went to water in the early evening, but even when she jumped a jackrabbit from under a bush she made only a token rush after it, just for the sport of it, and did not seriously exert herself to catch it. She was not hungry, and was content to putter around her hillside in the dark, idly noting the movements of other lives and listening to the sounds of the night. Ordinarily she would have traveled several miles by morning, to another resting place along her regularly patrolled circuit of the valley and its encompassing slopes. But there was more than one meal in that dead sheep, and that was reason enough to lay over. And so morning found her back under her tree, prepared to spend a second day in its shade.

By early afternoon she felt twinges of returning appetite, but she felt no compulsion to brave the heat in the flat, and not until the shadow of the hills was stealing across the valley did she set forth, reaching the gully just as the sun sank behind the western ridge. The carcass was just as she had left it, except that the birds had made further inroads, and a lingering but faint scent of man and of dog bade her be cautious. Her own parents had seized every opportunity to reinforce her innate distrust of man, but while she knew enough to keep out of his sight she had no personal knowledge of his malice. An older, more experienced coyote, having caught wind of the man scent, might have turned in its tracks and fled—although, as J. Frank Dobie remarked, the coyote's "unsleeping desire to investigate anything unusual often outpulls wolf caution." But although she was even warier in her approach than the day before, and scrutinized the desert at length to make sure no man was about, she let the utter stillness of the evening overcome her doubts and went to the carcass, by the same circuitous route she had followed the first time. She jumped into the gully at the identical spot, just as the man had hoped. There was a loud metallic snap, and something gripped her right foreleg in a sudden vicious bite.

Screaming in pain and terror, she leaped into the air to escape her assailant, only to be brutally jerked back to earth by a chain that fastened the trap to a buried stake. Again and again she leaped. Each time a searing pain shot up her leg to her shoulder, and each time she was yanked somersaulting to the ground. Finding she could not run from the enemy, she tried to attack it. She bit furiously at the trap, gnawed madly at the chain. But the terrible grip refused to relax. She dug at the ground with her free paw, wildly, aimlessly, snapping at chunks of earth that she rooted up. Soon shock and exhaustion temporarily dulled her hurt and fear. She had to rest, and she lay there panting, lifting her nose now and then to utter a forlorn, desperate, agonized howl. This sudden loss of freedom had her utterly baffled; nothing in her life had prepared her for this. One thing she understood, however, clearly if not logically. The man scent, the dead sheep, the trap, her suffering, all were related. And the common link between them was man. She had been taught as a pup to fear man, the enemy. Now she knew why.

Grim as her position was, the coyote had been fortunate in at least two respects. For one thing, it was evening. A coyote can survive exposure to the sun longer than, say, a rattlesnake, but had she sprung the trap twelve hours earlier and been bombarded all day, without shelter or water, by the full blast of the Southwest's July sun, she would have been dead long before dark. Also, the trap, when its jaws snapped shut, had damaged the principal nerves to her foot. The foot became numb, and after a while, having recovered from her initial panic and regained some of her strength, she began to chew methodically on her leg just below the steel jaws of the trap, where it did not hurt so much.

Shearing through her living tissue with her sharp-ridged carnassials, gnawing on her own bones with her powerful molars, she slowly amputated her imprisoned leg. Periodically she became impatient and renewed her attempts to wrench herself loose, only to calm down and resume her chewing. The ulna gave first. Then, as the sky began to pale imperceptibly in the east, there was a sharp crack: her teeth had crunched through the radius, and now only a thin strip of skin and flesh joined her leg and her foot. She quickly cut through this, too, and her foot fell away, severed above the wrist. Now was the time to try to break loose. Putting all her weight and strength into the effort, she gave a heroic tug. The jaws of the trap slipped off the stump, closing against each other with a snap. She was free—relieved of the trap, relieved of its cruel restraint, relieved, also, of her own right forepaw which, to his intense disgust, was all that the man found when he came by to check later in the morning.

The coyote was exhausted, yet she lost no time in fleeing from the scene of her torment. She headed east, halting every hundred yards or so to lick her bleeding stump, and limped and dragged herself to the top of the valley rim where she spent the day in the shade of a ponderosa. She continued her eastward travel in the two nights that followed, moving ever farther from her home valley and the man-threat it had revealed. The third night brought her to the edge of a long, flat valley that ran north-south between the mountain and a many-bayed wall of yellow-gray cliffs to the east. Here she turned south, hobbling along the western edge of the valley, until she reached the mouth of a small, rocky canyon cut into the mountain's flank. Finding that it offered a multitude of shelters and hiding places and a permanent seep of cool, clear water, she decided to call an end to her pain-wracked trek. Unbeknown to her, she could not have chosen a more ideal refuge. Some three miles back, where she had crawled under a fence, she had crossed the line between public grazing lands, which were supposedly dedicated to multiple use but where in effect the stockman was king, and a national park, where she and all other life forms enjoyed the protection of law.

Here, in a section of the park seldom visited by man, she holed up during the two months it took her wound to heal: two months of patient licking at the stump, keeping it clean of dirt and infection, two months of minimum activity and of minimum rations. Had she had a mate, he would have brought her food. A trapper in New Mexico even told me of a coyote he had found caught in a trap, with half a jackrabbit lying beside it—the meal, he said, must have been brought to it by its mate. But our coyote had to fend for herself, and by the end of summer, although thin and of limited stamina, she was healthy and ready to explore her new surroundings. She was a peg-leg, of course, but found the loss of a foot only a minor handicap.

It is often the crippled or decrepit predator, whether coyote or wolf, cougar or bobcat, who turns to killing livestock, just as in Africa or Asia it is usually the crippled or decrepit lion or tiger who takes to eating human flesh. For many of these animals it is a case of economic necessity: too lame to catch their natural prey, they are forced to depend for food on easily subdued domestic animals. For others, stock killing might even be a form of revenge against man, the author of their distress.

"Stories of three-legged coyotes are so plentiful," writes Joe Van Wormer in *The World of the Coyote*, "that one could almost assume that losing a leg bothers a coyote not at all. A female Michigan coyote had only stubs of front legs and looked like a kangaroo when she ran. At the time she was killed, she was carrying five unborn young."

An even more extraordinary case was described to me by Dr. Frederick Knowlton, who directs coyote research for the U.S. Fish and Wildlife Service, when I met him at Utah State University, in Logan. During a helicopter hunt, he said, two adult coyotes were spotted running from a den. One of them had a gait that was not quite normal: "No gross limp," Knowlton said, "just a little odd." The coyotes were shot, and when the men landed they discovered that the animal whose run had been a little peculiar had lost *both right legs*. "And it was still running!", Knowlton exclaimed, in a tone approaching awe. "Can you imagine what a balancing act that must have been?"

An Arizona Game and Fish biologist told me of a coyote whose lower jaw had been completely shot away. "When that poor guy was finally killed," he said, "the original wound had completely healed, so it had managed to survive quite some time without a lower jaw. How it ate at all, I've no idea. It was thin, in poor condition, but it was surviving. God, those animals are tough!"

The she-coyote had no such odds to overcome as did these other animals. Most of her prey were small burrowing rodents, who required for their capture a clever approach and a quick pounce, rather than a lengthy chase. Only the jackrabbits were generally too wary and too swift for her to run down unaided and, as it happened, she soon had help. Three nights in a row in the middle of November, as she coursed the valley near the mouth of her canyon, her serenade to the moon evoked answering howls from farther down the valley. This echoing voice sent a premonitory thrill through her body. Ever since she and her siblings had parted ways the previous autumn she had led a solitary existence. It was important that she learn to be independent and self-reliant. But the coyote, although not so gregarious as the wolf, is not by nature

a hermit as are the bear and the cougar. By now she was biologically and psychologically ready for company of her own kind,
particularly for male company. On the fourth night they met, in a
polite ceremony of nose-touching and tail wagging, the big dog
coyote all eager curiosity, the bitch masking her interest with
playful coyness and rebuffing his bolder advances with half-
annoyed nips. Introductions over, they hunted together for a while
that night, then each departed his separate way.

They repeated the process nightly for a week, meeting, hunting
jointly (on two occasions they teamed up to catch a jackrabbit,
which they shared), and parting again. Then early one morning,
when the bitch took off up canyon as usual, the male stood motionless, watching her go as if loath to lose sight of her. Twice he saw
her half turn and look over her shoulder in his direction. Diffident
but hopeful, he chose to interpret this as an invitation to follow.
She did not chase him off when he caught up with her but accepted
his presence as naturally as if they had been companions since
birth. From then on they slept together, played together, ran
together, and ate together. They found each other compatible: he
took no notice of her crippled right foreleg, there being nothing in
his coyote nature analogous to man's aesthetic abhorrence of
physical deformity. Each discovered that the other's company
added interest and zest to life, much as a domestic dog, if he is not
too jealous, will show renewed vivacity when a puppy is brought
into his household. Compatibility soon warmed to friendship. And
friendship grew into something deeper yet. Early in February the
bitch came into heat. On a cold, starry night the couple became
mates, which they would remain until death claimed one or the
other.

Their union was a choice, a decision freely made. Timber wolves
almost invariably mate for life, but evidence suggests that in coyote
society marital fidelity is less a law of the species than a matter of
individual commitment. As J. Frank Dobie put it: "In mating,
coyotes seem a good deal like human beings: some are strictly
monogamous and some are unstrictly polygamous. Some mated
males, like many professedly orthodox husbands, would apparently relish more variety and less responsibility. The brief season of
sex activity among coyotes no doubt retards promiscuity."

The female's sexual receptiveness died down in late February,
not to revive for another eleven months. But her mate did not
leave her to seek other conquests, or to resume bachelor life as the
males of many other species would have done. Instead, he became
increasingly attentive and helpful during the nine weeks of her

pregnancy. The loss of a paw made it harder for her to dig, and when she chose for her den an abandoned badger hole, burrowed into a south-facing talus across the canyon from the spring, it was the male who did most of the excavating to enlarge the hole to coyote proportions. It was he, too, who under her supervision dug two alternate dens, one a little farther up-canyon, the other a mile away at the edge of the valley. In the last month or so as the bitch's time approached, he relieved her more and more of the work of finding food; and now that the pups have come, and for several days after their birth, he does much of the hunting for two or, indirectly, for six. He is a fine young animal, with ample reserves of energy in his trim well-furred body, and his increased burden sits lightly upon him.

On returning from the hunt with freshly caught prey, perhaps a gopher or a rabbit, or a stomachful of field mice which he regurgitates, undigested, near the mouth of the den, he waits eagerly for his mate to appear in response to his bark, and waves his tail happily as she emerges from the den. Unlike some coyote fathers, he never enters the den himself, although he knows from their small voices that the pups are inside. The time soon will come when he will meet his offspring.

For the mother this is a brief interlude of rest and tranquillity, which enables her to gather the strength she will need during the demanding weeks of the pups' education. As she lies inside the den, having just nursed her babies, she watches their playful tumblings with a warm glow of affection. Her love for them is instinctive, undemanding, and uncomplicated. I will no doubt be accused of anthropomorphizing for ascribing such feelings as love and affection to a coyote. But I think it is dangerously presumptuous, indeed symptomatic of our sick hubris, to claim that such emotions are peculiarly human. I am convinced, by a lifetime of observations, that mammalian love is a psychic reality, and canine love, feline love, human love, are specific demonstrations and elaborations of a basic potential we all share in common.

Unlike the love of a human parent, the mother coyote's will not survive the break-up of the family: when the pups go their several ways, as they are likely to do in the fall, she will soon forget them as she prepares to bear another litter. Still, her attachment, while it lasts, is devoted, perceptive, and intelligent. She already can detect slight differences among the pups. One of them in particular calls attention to himself. He is the big one, the lighter-colored one, the first to have opened his eyes, the first to appear to recognize her as something more than a provider of milk. Her ears laid back, she good-humoredly allows him to worry her tail and nibble at her lip. She only protests when he attacks her crippled leg, stopping him with a short growl. Healed though it is, the stump remains tender, and every twinge of pain in it seems to awaken memories of those terrible hours in the trap.

She has no way of communicating to her inquisitive pup how she has lost her foot, so that he might learn about traps and how to avoid them. She can only, if the occasion arises, communicate to him her dread of man. This much nature has fitted her to do. It may be enough, if he is attentive and retentive, to improve his chances of survival in a world ruled by his enemies—and survival, after all, is what life is about.

In the week following the opening of his eyes, the big pup begins to feel a mounting restlessness. During the mother's ever-increasing absences, the big pup's explorations gradually reach beyond the confines of the inner den. Drawn to the light, he hazards up the tunnel, with his siblings, whose eyes have opened a day or two after his, scratching and scrambling along behind him. He stops short of the entrance, however. Out there, in the realm of light and undifferentiated sounds and smells, stretches the Great Unknown. Naturally curious, the pups are also naturally cautious, and they will not cross the threshold without permission and encouragement.

Permission finally comes on a warm, sunny morning when Big Pup and his fellows are nearly three weeks of age. They are hungry, and feeding time has come, but the mother has not made her usual mid-morning visit to the den. The pups stir about, whimpering, and as their impatience mounts they start scrambling up the tunnel. Just then they hear a soft whine, the voice of the father coyote, who is standing at the mouth of the den, cocking his head from side to side as he hears the scratchings of the pups inside. Again he whines. Although Big Pup has never heard this particular command before, his understanding of coyote language is innate and he knows that it means, "Come on out!" After but a momentary hesitation at the threshold he tumbles out into the sunshine, with his brother and sisters hard on his heels.

Exuberant at being released from the dark confinement of the den, Big Pup and his mates begin leaping about like four Mexican jumping beans. (In *God's Dog*, author Hope Ryden gives an eye-witness description of such a scene. This paragraph and the next are a close paraphrase of her account.) The pups move as if they have difficulty discovering how to activate their legs. When a voluntary impulse finally reaches the right muscle, it invariably is too strong, causing the pup to lurch forward in a most uncoordinated fashion and sometimes even to topple over. The father coyote watches them with keen attention, cocking his head at the excited squeals that accompany their clumsy capering.

Suddenly the mother coyote rushes down from her lookout post. The emergence of the pups is a momentous event for her, and she cavorts about her mate, until she is finally mobbed by the four babies. Then she holds very steady while each tiny creature stands up on its hind legs and takes hold of one of her teats. For five minutes she nurses them in this standing position, which allows her to keep an alert eye on her surroundings even while the pups are feeding.

His hunger sated, Big Pup stumbles about a bit, all senses a-quiver to the multiplicity of unfamiliar stimuli, then he flops down in the shade of a neighboring limestone boulder. He sees the father coyote come towards him, tail gently waving; awed by the approach of this large being, so like his mother and yet somehow different, he adopts a submissive position, rolling onto his back and exposing his creamy belly. He feels the big dog coyote sniff him all over and give his face a brief lick; at first he submits to this investigation without moving, but as apprehension graduates to pleasure he begins to wriggle under the exploratory nose, and when the father coyote straightens up and moves away to make the acquaintance of another of his offspring, Big Pup gets up to follow him, only to stumble in an entanglement of paws and sprawl in the dust. The warmth of the sun, the warmth of his mother's milk inside him, act as powerful soporifics. In a second the pup is asleep.

Sleep, eat, play, explore—and learn. Such is the program of activities for Big Pup and his littermates from their first day out in the open. Sleeping, at night, is still done in the den, out of reach of the stealthy powers of darkness—the soft-footed bobcat and the silent-winged great-horned owl. But daytime naps more and more commonly are taken outside, in the warmth of the morning sun or in the shade of a rock or shrub, and always under the vigilant eye of one of the parent coyotes. Eating takes place four times a day, at four-hour intervals, with the mother serving as a stand-up lunch counter.

Play and exploration occupy the rest of the days, with both increasing in variety, exuberance, and adventurousness. At first the games are a clumsy rough-and-tumble, more tumble than rough. But as the pups gain in strength and coordination the patterns of play become more complicated. There is not only general rolling around in the dust, but also games of stalk and pounce, chase and counter-chase, nip and flight. These games are not merely an outlet for excess energy: They are a learning experience in themselves. One thing Big Pup learns, besides how to control his limbs, is to control the strength of his bite. By the third week after their emergence from the den the pups have begun to teethe. One afternoon, in the midst of a playful tussle, Big Pup chomps down on his brother's tail; the brother lets out a squeal, whirls around and closes his little teeth on Big Pup's ear; now it is Big Pup's turn to squeal and a real scuffle starts, punctuated by squeaks and high-pitched growls. The fray finally is broken up by the father coyote who trots up, grabs Big Pup by the scruff of the neck, and sets him down some distance away from his antagonist.

These scraps become more common as the days go by. Sometimes they start in a dispute over a parent's attention. Sometimes they are set off when one pup mounts another's back and the latter objects to this attempt to assert dominance. Both the spats and the back-mounting are part of the learning process: they serve to determine the rank which each pup will hold in the eventual hierarchy of the litter. Very early in these contests Big Pup manages to establish his primacy over his peers. He is not a bully, but he proves himself a little more vigorous, a little more assertive, a little more determined than his fellows, and soon they begin to defer to him. Once the pecking order is established the squabbles become less intense: a threat gesture is usually enough to make a subservient pup back down, without the confrontation having to be resolved by violence. The pups are becoming socialized; and the protocol they are learning will be vitally important when, as adults, they must interact with others of their kind, such as in defending a territory without having to kill or be killed.

The parents tolerate these mélées with little interference unless the loser's protests become too shrill, in which case the father or mother breaks it up. The parents, furthermore, are enormously patient in absorbing the abuse heaped on them by their little ones. The pups delight in crawling all over the adults, in nipping their tails, in ganging up on them. The parents take an active part in the play—a characteristic of coyotes and other highly intelligent, social animals. Once in a while, nipped a little too smartly by a pup's needlelike teeth, the mother coyote retaliates with a mild snap. The father, when he can take no more, simply gets up and trots off at a pace which the little ones cannot maintain. Only when one of his pups gets confused and tries to nurse him does the father jettison his usual sober dignity; on such occasions he leaps up with a look of outrage and takes off down canyon at a gallop.

The budding of puppy teeth precipitates a major change in the litter's life-style. Nursing, for the mother coyote, is beginning to be a painful process: every so often a pup, entranced by the bliss of suckling, closes his sharp little incisors on the nipple, which causes the mother to yelp and to jump away. The time has come, she realizes, for another big step in the pups' education: they must be weaned, introduced to solid food, taught to accept the staples of a coyote's diet. The mother takes the initiative in the process, with the later assistance of her mate. At feeding time one morning, she comes trotting up the canyon from the valley, where she has been busy catching mice since first light. The pups see her coming and bound forward to greet her, eagerly anticipating their usual breakfast. To their surprise and consternation, instead of standing still for them and presenting her flank, the mother flattens her ears and faces them with the fearsome coyote gape—a wide-open-mouthed grimace that shows off every one of her forty-two formidable teeth. Before the pups have a chance to recover from their shock, the mother begins a rhythmical heaving and after a few spasms coughs up a slimy mess of premasticated, semidigested field mice. Then she lopes up to the look-out station to join her mate.

The pups are nonplussed by this performance. They have seen their father disgorge food for their mother, and watched her devour it hungrily. Have the roles been reversed? They look to the sentry post, half expecting the father coyote to come down and eat up the offering. But he just sits up there motionless, looking down at them; their mother lies beside him, her eyes also fixed on them. As usual, it is Big Pup who proves the most enterprising. He waddles uncertainly up to the mice and sniffs them over dubiously. Not pup food, he decides. Still it smells interesting—he remembers his mother's breath in the den—and with a sharp snarl at his suddenly emboldened siblings that makes them back away, he flops down on top of the mess and luxuriously rolls about in it until his fur is thoroughly coated with its juices. Then he stands up, shakes himself, and remembers his hunger. Looking up at the mother coyote he begins to whine and whimper, half plaintively, half petulantly, with his brother and sisters joining in chorus.

Finally, after an hour, she condescends to come down, allows the pups to nurse briefly, and breaks away at the first nip of a puppy tooth. And then she herself gobbles up the spurned breakfast. In the early afternoon she humors them with another brief nursing, and gives them a thorough grooming—especially the redolent Big Pup. But she ignores their hungry cries when sundown arrives and stands at the lookout station peering intently down canyon. Presently the father coyote comes trotting up from the valley. He pauses briefly to touch noses and acknowledge his mate's effusive greeting with a wave or two of his tail, then climbs down to where the pups are waiting on the patch of bare earth in front of the den. With a few heaves he repeats the mother's performance of that morning.

By now Big Pup and his mates are famished. Again, as in the morning, Big Pup takes the lead, goes to the mouse meat and smells it carefully. Unlike this morning, he does not roll in it. Instead, he gives the mess a few tentative licks. It tastes stronger, richer, less sweet than his mother's milk. But he likes it! He takes a small bite; it slides down his throat without gagging. His siblings have been watching him closely, noting his every move and his every reaction. They now move in, too, ignoring his growl, and begin to sniff the meat. This spurs Big Pup on to eat more rapidly, lest he get less than his share; his appetite is contagious, the others begin to eat, too, and soon they have wolfed down the entire meal. Thus auspiciously started, the weaning process takes a week, during which meals of solid food are supplemented once or twice a day by a meal of milk. At the end of the week the mother's milk dries up; from now on and for the rest of their lives meat will be the staple of the pups' diet, with occasional vegetable supplements. And with this change of diet the pups' chances of survival are automatically doubled. For up to this point they have been utterly dependent on their mother's supply of milk. If she had died they would have died also. Now, even if something were to befall the mother coyote, the father could and would see to it that the pups were fed.

Yet for one of the pups, the rest of its life consists of three days. From the very first day outside the den, the little ones have spent much of their time investigating their surroundings. The explorations at first were short and timid, extending to the limit of the open area outside the den, which is bounded by a big limestone boulder, a Gambel oak, some cliff rose bushes, and a few tall spikes of yucca. As their eyesight sharpened the pups could look out into this mysterious greater world—they could see the sandstone cliff face across the canyon, with its streaks of desert varnish; they could look past the mouth of the canyon, a hundred yards away, to a portion of the flat green expanse of the valley. But in the first two weeks any attempt to wander away from the playground brought a parent trotting down from the lookout station to herd the little explorer back to the fold. The pups learned very early to heed their parents' commands—especially the short, downward sliding howl which sent them scurrying back to the den. Another urgent alarm which the pups instinctively heeded was the cry of the ravens, those big black birds who roosted in crannies of the cliff and whose aerobatics the pups delighted in watching.

But as time went on, the pups were allowed to extend their radius of play and investigation. They soon discovered the seep of water across the canyon, and now that they are weaned this has become their favorite haunt. The trickle of cool water tastes good, and they never tire of padding around in the patch of moist earth beneath the spring, or of watching the darting, hovering flight of the red dragonflies and blue damselflies that frequent the spot. It is the movement, not the colors, that attracts them: like all members of the dog family, the pups are color-blind, their world a kaleidoscope of black, white, and shades of gray. The activities of other creatures now intrigue the pups, and have awakened in them embryonic hunting instincts. It is more fun, they discover, to stalk and pounce on an elusive butterfly than on a brother or sister.

One day Big Pup notices a largish insect with a furry white abdomen scurrying over the ground. He chases it, slaps a paw down on it, and jumps back with a loud yip of surprise and dismay: this is a "velvet ant," a wingless wasp with a vicious stinger which she has buried in a pad of his paw. Big Pup will remember the lesson and leave these disagreeable creatures alone in the future. But a few days later, as he returns with a sister from an hour's digging near the spring, his path crosses that of another new insect, a large black beetle that is unhurriedly but purposefully marching down canyon. Big Pup alters course to trail the shiny armored arthropod, his nose an inch from its tail, and his sister follows just behind him. They are fifty yards down canyon from the den before the father coyote, drowsing at the lookout station, realizes they have left the vicinity of the spring and sets off to retrieve them. But another, far swifter than he, has spotted them first. The attack is over so quickly that neither pup has a chance to realize what is happening. The little female feels a powerful jolt, a quick stab of pain, and then nothing—the golden eagle's talon has pierced her heart, and she is already dead as he lifts her away. Big Pup feels the rush of air as the eagle brakes its vertiginous dive, hears his sister's stifled shriek, and ducks as a large shadow passes over him. When he picks himself up and looks around, he sees only his father racing towards him.

Even the father coyote, although he witnessed the termination of the eagle's swoop and saw the huge bird flap away down canyon with something large in its talons, is not quite sure what has happened. When he reaches Big Pup he noses him over to make sure that he is well, sniffs suspiciously at a drop of fresh blood on the sand marking the spot where the girl pup took her last step, then shepherds Big Pup back to the den area. The male coyote is uneasy; his uneasiness communicates itself to the three remaining pups and to the mother coyote when she trots up a short while later. She investigates the little ones thoroughly, and gives them a careful grooming. But although aware that something is amiss she does not specifically realize that one female pup is absent. Either parent coyote would have grieved at the loss of its mate. But the vanished pup, one out of four, is not personally missed either by her parents or her siblings. Hope Ryden, in *God's Dog*, conjectures that coyotes may have a sense of number, or at least a sense of quantity. But these coyotes' arithmetic evidently is insecure, as it proves to be again the next day.

In the late afternoon the mother coyote trots out onto the valley floor, pausing now and then to investigate the opening of a ground squirrel's burrow. A sudden shift of wind makes her stop in her tracks. After several months in the area she has become fairly inured to the sight of an automobile three miles away across the valley, as it creeps along the dusty road past the ranger's house on its way to or from a scenic overlook five miles to the south. And although her hackles still rise, she no longer panics at the faint human scent which a rare easterly breeze sometimes wafts to her nostrils. This time, however, the human odor is much stronger, and from the depth of her unconscious there wells up a memory which, while it may not resolve itself into a precise association of man, steel trap, pain, loss of foot, still fills her being with an elemental fear. She pauses long enough for her sharp eyes to spot two small human figures on foot; they are not on the road or by the ranger station where they belong, but halfway across the valley and they are coming towards her. As soon as she has established this, she whirls around and races up the canyon at top speed which, even minus a foot, is considerable. Near the den area she utters the warning signal in a tone so peremptory that the pups scurry into the den. The mother coyote dives in after them, hustles them back to the inner chamber, and turns around and scrambles back almost to the entrance, so that she can look out but not be seen.

The father coyote stands tensely at the lookout post, all his senses straining to detect what it is that has so upset her. "A feather fell from the sky," an Indian saying goes. "The eagle saw it, the deer heard it, the bear smelled it, the coyote did all three." Presently his nose catches a first whiff of man scent, and he tenses up still more. Almost simultaneously his ears distinguish the sound of human voices from the raven's alarm cry and the cooing of a pair of doves. As both voices and scent grow stronger, the father coyote crouches motionless at his post, torn between the urge to flee and the desire to stand by his family. He determines on a compromise. When at last he spots the two humans entering the mouth of the canyon he jumps to his feet and quickly climbs the short slope above his lookout to the foot of the canyon's north wall. There, in a patch of shade, he lies down and without moving a muscle other than those that work his eyes, ears, and nose, he watches the human couple trudge up the sandy wash.

"There it is, just where the map says," the man's voice rings out. "And look, Ginny, there's a trickle of water!" "Oh, what a lovely spot!", the girl's voice answers. "Why couldn't we camp right here?" "No, honey, no," the man laughs. "We'll have a drink and take some pictures, but we've only come three miles, and there's still three hours of daylight. We've got to push on." They walk up to the spring, shrug off their packs, and fill their cups from the seep. They stand barely a hundred feet from the mouth of the den, a hundred and fifty feet across the canyon from the father coyote, who now makes his move. He stands up, lets out three sharp barks that crack like pistol shots and cause the couple to whirl around and face in his direction, then he leaps out of the shade in full view of the intruders and gallops down the canyon. "Look, Dave, a *coyote!*" the girl calls out excitedly. "Oh, marvelous!" the man cries. "And isn't he a beauty? Bet he was having a drink when he heard us come up. We must have scared the poor guy to death." "You don't suppose there could be a den somewhere?" Ginny asks. "What luck if we could see some pups!" "Yeah," says Dave, "but let's not go looking for them. We might spook the parents so they'd abandon the young. I don't know how coyote parents react." "If there *is* a den around here, could that coyote be trying to decoy us away?" "Could be. So let's push on and not make him any more nervous than he is."

All this time the mother coyote, lying frozen with apprehension just inside the dark mouth of her den, keeps her piercing yellow eyes fixed unblinking on this human pair. Big Pup and his mates feel the contagion of her fright. Besides it is stuffy and crowded down in the den, and Big Pup at one point begins to whimper softly. The mother coyote cuts this short with a barely audible, but no-nonsense growl. Her fear, hardly allayed by the fact that these humans speak in gentle voices and move about quietly, becomes almost intolerable when, having re-shouldered their packs, they take several steps away from the spring directly towards her. It seems to her for one agonizing second that the man's eyes are looking directly into hers. But then he turns to his left, as does the girl, and the sound of their boots crunching along the sand gradually grows fainter as they move away.

As soon as they are out of sight, the father coyote trots back to the den. He pauses at the entrance, looks into his mate's face, then walks a short distance up canyon, sniffing the hikers' trail. He stops, his eyes fixed on the bend in the canyon from beyond which faint voices can still be heard. A half hour later the voices have disappeared into the distance and he trots back to the den. Looking down at his mate he whines and waves his tail. She is calmer now, calm enough to come out into the open. But when Big Pup and his fellows try to follow her out she gives them a sharp command to stay.

Big Pup, who is nearest the exit, obeys grudgingly. Now and then he begins to whimper from combined boredom and hunger, but there will be no dinner this evening, until much later. Instead he is in for an upheaval which will reinforce the impression of anxiety communicated to him during the whole time the humans were near. Not having had any experience with man before this, Big Pup does not, and may never, share his mother's feeling of near-panic at the threat of human contact. But he has understood her message clearly: when man is near, beware!

Coyotes' faces and bodies are marvelously expressive of their moods and intentions. There is a semaphoric code, involving the attitude of the ears, the curl of the lip or the gape of the mouth, the port of the head and of the tail, the character of the walk—stiff-legged, relaxed, slouching—immediately intelligible to all coyotes and almost as much so to the empathetic human. Coyote voices, too, have an enormous repertoire of sounds, from whines to growls to yips to barks to howls and modulations and combinations of these, that certainly carry meaning, and perhaps complex messages, to listening coyotes. Many men have claimed to understand coyote talk, believing a certain vocalization signals the approach of enemies, or warns of impending death, or gives notice of forthcoming change in the weather. There is a saying in northern Mexico: *Cuando llora el coyote, se va a llover; cuando grita, se seca*—when the coyote wails, it is going to rain; when he shouts, it will be dry.

Then again, there are times when coyotes' actions make clear that a specific, precisely inflected signal has been exchanged and understood, although the signal itself is far too subtle to be detected by even the most acute human observer.

Just such a signal now passes between the mother and father coyotes. Each of them understands perfectly what is proposed to be done, when, and by whom. Shortly after sundown the coyote mother walks to the mouth of the den and whines once. Big Pup tumbles out, anxious for a romp, but not a chance. She grabs him by a hind leg, lifts him off the ground, and carries her wriggling charge down canyon. At the canyon mouth she turns right, and ten minutes later Big Pup finds himself deposited gently on a mound of soft earth. He picks himself up, shakes himself, and looks around. He is out in the open, a few feet up the western mountain from the level valley floor. A piñon pine looms over him, and at the foot of the tree yawns the opening of a den into which the mother now crawls. She calls him to follow her, then brushes past him and disappears out into the dusk.

This is the first time in his life that Big Pup has been left all alone, and he is frightened. He almost decides to try to follow his mother but stops at the den entrance, hesitant in his new surroundings. Soon he detects the sound of soft footfalls, and in the half light he recognizes his father and mother, each with a pup hanging limply from its mouth. Overwhelmed with relief, Big Pup forgets the admonition to stay put in the den and barrels out to meet them, voicing his joy with a torrent of squeaky barks and howls. And for once his parents overlook his disobedience: they themselves feel intense relief that the human threat has passed and that the move to the new den has been completed without trouble.

Except—they seem unsure. The father coyote, not quite at ease, takes off. He trots back to the abandoned den, looks into the entrance, cocks his head to listen for puppy sounds within, and, hearing nothing, goes inside. Satisfied that the den is indeed empty he retraces his steps and rejoins his family. Then, as if she mistrusts the thoroughness of his search, the mother coyote runs back to the old den to make an inspection of her own. Are the parents unable to count their three pups? Or could it be that their unconscious remembers that once there were four? At any rate, not until the mother finally returns to the new den does the coyote family entirely relax.

A translucent darkness now fills the valley; the long scalloped wall of cliffs that encloses it to the east emits a ghostly reflection of the sunset's afterglow. The air is still and soft, lightly scented with sage. The day birds have settled down for the night; a few sleepy voices murmur from the scattered junipers. Now the poor-wills begin their monotonous calling—a tuneless, wistful, haunting sound. The coyotes' sensitive ears pick out the tiny chirps and squeaks of countless little rodents astir in the valley grass. Dinner is out there waiting, and soon the parent coyotes will go forth and fetch it for themselves and for the pups. But first, as is the custom of coyotes everywhere, they must announce their contentment.

Pointing their muzzles to the sky, they break into song. This is the coyote "Ode to Joy," the quintessential voice of the desert, the jubilant carol that made Dobie's vaquero laugh with empathy and call out, "¡Cantad, amigos!" In this primal cantata, the pups' uncertain, falsetto notes play counterpoint and syncopation to the parents' long, controlled tremolos. When the voices die down, an answering refrain comes muted but clear from the opposite side of the valley; then a group far up-valley picks up the theme, and then yet another well up among the ponderosas of the western mountain. Other coyote families, perhaps with reasons of their own to be glad, perhaps only from an urge to proclaim their common coyoteness, are helping to fill the night with exuberant song. In their camp halfway up the mountain the man prods his sleeping companion. "Ginny, wake up! Listen!", he whispers urgently. The girl stirs in her sleeping bag, props herself up on her elbows, and stares out over the valley. As the last chorus ebbs away she breathes: "Dave, I'll always remember this as the night of the coyotes!"

Big Pup and his fellows adapt fast to their new surroundings, and find them very much to their liking. They delight in the openness of the valley, the unimpeded prospect, the spaciousness of a landscape that beckons them to dig and to roll and to race and to chase where they will—subject only to their parents' surveillance. They experience a new feeling of freedom to which they respond with increased exuberance and daring.

Evolution made the coyote an animal of the wide open spaces. Before the white man spread across the continent with his flocks and his herds, his guns and his traps, his technology and his cities, the coyote was essentially and almost exclusively an animal of the broad-vistaed reaches of the trans-Mississippi West. "Prairie wolf" or "brush wolf" were early terms for him, contrasting him to the larger "timber wolf."

Human intrusion and human enmity forced the coyote to adapt so that he now occupies territory that once was alien to him and, ironically, by clearing the forests and exterminating his competitor the wolf, man has inadvertently favored a vast extension of the coyote's original range. Now the little "prairie wolf" is at home from the forests of Maine to the city parks of Los Angeles, from the tundra of Alaska to the mountains of Guatemala. But still he remains, *par excellence*, an animal of prairie, basin, and bajada, rather than of forest, range, or rimrock. And so, Big Pup feels an atavistic compatibility with his new surroundings that he unconsciously missed in the confines of the natal canyon.

He is now two months old. His coat has lightened to a fulvous grey, with just a hint of the rust, black, and white accents that brighten the coats of his parents. The black-tipped guard hairs have yet to show through the soft underfur, but his muzzle and ears have lengthened; he is outgrowing the bundle-of-fur puppy look and is becoming recognizable as a coyote.

He is a long way, however, from outgrowing puppy insouciance and the constant need to explore and try new things. Big Pup is now quite able to devise his own amusements, and part of every day he spends by himself, chasing insects or just poking around, a necessary preparation for independent life ahead. He uses these solitary periods to familiarize himself with every detail of his surroundings. He finds that sagebrush smells sweet but is bitter to the taste, that it is delicious to roll on the carcass of a long-dead skunk but disastrous to challenge a live one.

His parents take an active role in broadening his experiences. Just a week after the move to the new den, the father coyote returns from the hunt one morning with a still-warm jackrabbit in his mouth and deposits it in front of the three curious pups. Big Pup immediately pounces on this splendid new toy, grabs it by a leg and dares his brother and sister to try to snatch it away from him. A glorious tug-of-war ensues, punctuated by mock-ferocious growls. As they tighten their holds on the rabbit, the pups' teeth penetrate its hide, and fresh warm blood spills out, activating the pups' taste buds and teaching them that this is more than a toy—it is food. The already tattered rabbit soon is dismembered, and each pup walks away with the portion it has scavenged to feed on it at peace.

The step is short between eating whole dead prey and catching live prey. In fact, the pups already are hunters: from merely pouncing or snapping at crawling or flying insects they have graduated to catching and eating them. Big Pup in particular has become expert at seizing grasshoppers, whose taste he relishes. So expert, in fact, that he has taken a step towards insuring his survival just as crucial as the one he took when he allowed himself to be weaned. If he had to, he now *could* feed himself on these captured insects and the sweet juniper berries.

Indeed, such foods, at certain times and places, can be the major or exclusive component of a coyote's diet. I have seen coyote scats that consisted almost entirely of partly digested juniper berries. An unpublished study by Arizona State University's Drs. Robert D. Ohmart and Bertin W. Anderson found that coyotes' diets along the lower Colorado River could be primarily vegetarian; in the fall of 1974, honey mesquite and screwbean mesquite pods formed four-fifths of the bulk of their scats.

Ever adaptable in his tastes and his ways, the coyote may have food preferences but he has few dietary taboos. "Among the objects that coyotes are known to have swallowed, either whole or in part," J. Frank Dobie writes, "are pieces of string, paper and cloth, rubber from an automobile tire, harness buckles, snails, beetles, horned frogs, wildcat, house cat, skunk, armadillo, peccary, grizzly bear carcass, eggs of birds, turkeys, turtles, grasshoppers, crickets, any kind of mice, rats, all other rodents, a great variety of wild berries, grapes, dates, peaches, prunes, carrots, sweet peppers, tomatoes, watermelons, plums, pumpkins, oranges, tangerines, bumblebees, flies, beaver, crayfish, any kind of fish in reach, bull snakes, rattlesnakes, other snakes, centipedes, apples, acorns, pears, figs, apricots, cherries, cantaloupes, porcupines, ants, coyote meat and hair, all sorts of water birds, all sorts of land birds, including the turkey vulture, pine nuts, peanuts, grass, frogs, honey, green corn, bread, sugar and spice and everything nice, new or old, hot or cold, cooked or raw."

The pups have now begun to tag along with their parents on short hunting forays out on the flat below the den. They observe and try to emulate their parents' techniques: the stiff-legged, four-footed mouse-pounce; the motionless point, and sudden rush after an antelope ground squirrel; the cooperative jackrabbit hunt in which the adults relay one another as they pursue the quarry in gradually tightening circles, and in which the climax often comes when one adult, hiding behind a sagebrush plant by which it knows the other will drive the tiring rabbit, dashes out at the last second and grabs the victim as it goes by.

The day comes when Big Pup makes his first successful pounce on a kangaroo rat; triumphantly he swaggers before his father and siblings, the trophy dangling from his mouth, only to lose it to his brother who makes a lightning-quick grab for it. On his first squirrel hunt Big Pup succeeds in making contact with his intended prey, but instead of his biting *it*, the squirrel bites *him*,

closing its sharp incisors in his lip. He yelps and shakes his head, sending the squirrel flying, but smarting from the bite, and with the taste of his own blood on his tongue, he does not pursue it.

Learning has its price, and sometimes the price is high. The swift-darting lizards that race around the old lava flow just south of the den are a particularly frustrating quarry: never yet has a pup managed to catch one, although the girl pup once did succeed in seizing one by the tail, only to see the rest of the lizard scurry away.

Then one morning, not far from the lava flow, Big Pup and his brother come across a novel lizard which sends their hopes of success soaring. It is much bigger than the ones they have been used to chasing, but it is also slower for it seems to have lost its legs; it does not scurry like the others but crawls along on its belly. In fact when they run up to it, it does not even try to flee but winds itself into a coil and faces them, its triangular head raised a foot from the ground; it holds its tail erect and vibrates it visibly with a churring sound that is new to the pups' ears. Taken back by this inappropriate lizard behavior, the pups halt a couple of feet from it and begin circling it warily while expressing their mixed emotions in a stream of excited barks. At one point, while his brother holds the rattler's attention, Big Pup makes a feint towards it; the snake apparently has expected this for it turns and strikes at Big Pup, who leaps back just in time. Then his brother lunges, but he is not so lucky. The snake's fangs close for a second on his cheek, the pup lets out a scream of pain and tumbles backwards, shaking his head.

Alerted by the commotion, the father coyote runs up, and when he sees the cause of it his hackles rise and he gives the pups the sharpest command they have ever heard, then begins to dance around the snake, always careful to remain just beyond striking reach. Again and again the snake strikes and misses; again and again, before it can re-coil, the father springs forward and snaps it just back of the head. The sixth time this happens the snake collapses. The father coyote seizes it in his teeth and shakes it violently until it stops squirming. Finally he bites off the head and carries the body to the vicinity of the den, where he eats it.

The pups follow him back, but the bitten pup stumbles twice along the way. He feels weak and dizzy, and when he makes it back to the den area he flops in the shade of a basalt boulder and refuses to stir from the spot. By early afternoon the pup is very ill. His head is badly swollen, it throbs violently and he whimpers at the pain, his pulse is rapid and weak, and he responds listlessly when his worried mother comes over to sniff and to groom him. She lies down a few feet from him, watching him steadily, and warns away the other pups when they want to tease him into play. After a little while the whimpering ceases; the pup's breathing becomes rapid and shallow; excess saliva drools from his chops. A little later, when his mother goes over to nudge him, he responds not at all; shock and paralysis are setting in, and at midafternoon his breathing stops. The mother coyote notices this at once; again she gets up and goes over to him, sniffs him searchingly, nudges him a little roughly as if to force him to stir, then accepts the inescapable: her pup is dead. She picks up his body in her mouth and carries it off a half mile from the den and deposits it under a large sagebrush. As it happens the next morning, in the course of his daybreak hunt, the father coyote's nose leads him to the dead pup. He recognizes the corpse by scent as that of his offspring; yet to him it is now only another dead creature, a piece of carrion fortunately encountered, and he promptly proceeds to devour it. Coyotes, ever practical, have no revulsion at eating the dead of their own kind.

With the litter now reduced to two, the coyote parents find their work load lightened, especially as Big Pup and his sister have begun to help provide for themselves. The pups' hunting skills and successes are gradually increasing. But hunting to them is still more of a sport than part of the serious business of earning a living. Once in a while one or the other, upon sighting small prey, will freeze and hold a perfect point for minutes on end, but just as often their concentration can be broken by a fluttering butterfly, and they will break the stalk to snap at the bright-winged insect. Of course the parents themselves are frequently happy to interrupt their mundane pursuits and engage in a bit of fun, especially at the expense of creatures devoid of their sense of humor.

One July afternoon, towering thunderheads pile up over the shoulder of the western mountain, presaging the season's first storm and breaking the heat of the sun. The air is charged with tension and a feeling of impending violence, which stills the songs of the birds but only quickens the coyotes' nerves. The four are trotting along the slope some distance above the den, to no partic- ular purpose except perhaps to work off excess energy, when all of a sudden Big Pup and his sister, bringing up the rear, see their parents bound ahead at full speed yodeling the hysterical yip-yip- yip-yip hunting cry. The pups, rapidly left behind, follow their parents' retreating forms and voices as best they can. Presently they notice that the parents have halted their headlong chase and now are running circles around a big limestone boulder, still bark- ing excitedly. When the pups rush up, panting, they find that standing atop the boulder is a large and very angry male bobcat. His back arched, his ears laid back, the fur of his ruff and back erected like the spines of a hedgehog cactus, the old tom spits and snarls in frenzied outrage. A bit scared to approach too closely, Big Pup and his sister stand off, yammering their encouragement, as the parents continue prancing around the boulder, their taunting barks perhaps nature's closest equivalent to human laughter at slapstick comedy. Indeed, it is difficult to escape the conclusion that coyotes *do* have a sense of humor. How else to explain, for instance, the well-known propensity of experienced coyotes to dig up traps, turn them over, and urinate or defecate on them?

The father and mother coyote have no desire, at least on this afternoon, to try to kill the bobcat, even if they could ("It takes two darn good coyotes to stretch out a big old tom," Dick Randall said to me), but they are hugely enjoying his discomfiture, and they only break up the party when an apocalyptic thunderclap, which frightens the pups half to death and sends them scampering back to the shelter of the den, announces the imminent start of the deluge.

The den is still a haven in emergencies such as this thunderstorm. But otherwise the pups, now three months old, use it less and less. They now prefer to sleep outdoors, and as summer wears on and they accompany the parents on longer and longer excursions through their territory, the nights begin to multiply when they do not return to the den area at all but bed down wherever they happen to be at the end of the evening hunt. Big Pup and his sister are beginning to learn that their world does have boundaries, imposed not so much by topography as by the unwritten pacts of coyote society. Their family's territorial limits are the fence line marking the northern boundary of this national park, the dirt road running down the eastern side of the valley, and an ill-defined line midway up the slope of the western mountain. The land to the south is at present unclaimed, its two resident coyotes having died earlier this year. Twice Big Pup is present when territorial disputes are adjudicated. On one occasion he sees his father chase a male coyote back across the road; there is no actual fighting, and the pursuit stops as soon as the intruder regains his own ground. Another time Big Pup watches as his mother is chased and overtaken east of the road by the resident female, and allowed to regain her own turf only after a humiliating display of submission.

As summer eases into the cooler, shorter days of fall, a difference in temperament becomes increasingly pronounced between Big Pup and his sister. She is content to tag along with her parents; where they hunt she hunts, where they sleep she sleeps. She has another year to go before she can experience the urge to find a mate, and she feels no desire to break the parental tie. It seems likely that she will remain with the father and mother coyotes through the winter, and that if her mother bears another litter in the spring she will share with her parents the baby-sitting duties and the chore of finding food for the new pups.

Big Pup, on the other hand, is increasingly restless. In his first six months he has learned all the essentials for coyote survival in his particular environment. Except that several months must yet elapse before he gains his full adult weight of thirty-odd pounds, he is in outward appearance a full grown coyote. As the weather cools his fur grows thicker, and there are handsome color contrasts between the reddish tinge of his ears and legs, his black-flecked fulvous body and black-tipped tail, and his creamy white underparts. The bonds of affection between him and his father and mother and sister are still strong, but he finds himself spending more and more time alone in the vacant territory just south of his parents' home range. At first he explores it by himself during the day, returning to the family camp at night. Then he takes to staying away two or three days, then a week, at a time.

By winter's first frosts Big Pup has made the break. He has staked out the southwestern portion of the valley as his own and regularly patrols its perimeter marking scent posts with his urine. Every now and then his and his family's circuits happen to converge, and there is a joyful ceremony of nose-touching and tail-waving and perhaps a joint hunt. But the rest of the time Big Pup minds his own business. He faces none of the rigors coyotes must endure in the northern states, where blizzard may follow blizzard and starvation stalks his kind. At the valley's 4,700-foot elevation the snowfall is light and soon melted. Food is adequate, if not so plentiful as in the spring or summer, and Big Pup weathers his first winter without notable incident or difficulty.

It is just over a year, at this writing, since Big Pup took his first breath in the darkness of the den: a tiny, blind, helpless being utterly dependent on his parents' care and devotion. In the past twelve months he has grown to be a strong, quick-witted, self-reliant young animal, with his species' combined toughness and zest for life. Though he lives by himself, he is no morose loner like the badger or cougar or black bear. His is a blithe spirit, and for all his caution and suspicion, he takes delight in his activities. Big Pup has learned much in these twelve months, and he will continue to learn throughout his life, for the coyote has a native capacity for experiential learning. Ever alert to the slightest change in his surroundings he registers all, remembers, and adapts.

In reaching his first birthday, Big Pup has survived the period during which a coyote is most vulnerable. Of course, in most parts of the coyote's western range, the second and third years are almost equally perilous, and the peril comes from man. Fish and Wildlife researchers, trying to discover patterns of natural mortality in a coyote population, have been thoroughly frustrated.

Dr. Frederick Knowlton, the coyote research director, told me: "We know there *has* to be a natural mortality. Some of them *must* die of disease or accidents or just plain old age. But we can't *find* them. All we find are those that were killed by man—traps, bullets, road kills. If a coyote reaches the age of four, then he has a pretty good chance to live to old age."

Had Big Pup been born in most other parts of Arizona or New Mexico, he still would have to weather two or three years of trials before he could hope to reach his biological life expectancy of ten to fifteen years. Had he been born in New Mexico in the sheep country around Roswell, or the sweeping cattle country around the Plains of San Agustín, he might become a lamb- or calf-killer, but whether he did or not would have little bearing on his fate: the mere fact of being a coyote would make him an outlaw. And some fine morning his homeward trot could be halted abruptly by the snap of a trap, which he might not have the luck to escape as his mother did. Or, seeking to squirm under a pasture's net wire fence, he might feel something grab him around the neck, and tighten relentlessly as he tries to pull free till he dies of asphyxiation, garroted by a snare. Or he might be running a jackrabbit across a flat near the Rio Salado, and suddenly hear a queer clacking sound approaching from behind; looking up, he would see a giant, bubble-eyed dragonfly bearing down upon him; he would take off at his full speed of forty miles an hour or better, but his speed would not save him; there would be a sharp crack, he would feel the buckshot slam into him, and the rest would be silence.

He could have been born in the Kaibab National Forest, near Flagstaff, Arizona, where part of his diet might consist of deer or antelope fawns, or on the grass-and-oak savannah west of the Santa Rita mountains near Tucson, and fall prey to deer- or javelina-hunters who would see in him a competitor for "their" game. Again, he might have been born at the base of Camelback Mountain in Phoenix or in the foothills of Tucson's Santa Catalinas, and be duped by the phony rabbit cry of a varmint caller and shot, or, dazzled by the headlights of a speeding car, be smeared across the asphalt by a screeching tire.

In all these areas many coyotes Big Pup's age do survive to raise families, and some even attain old age. But these are most often either the very smart, who have learned, often painfully, the ways of man, or the very lucky. Of course Big Pup, his sister and parents are very lucky in their own way: they live well inside the haven of a national park, where their right to go about their business unmolested by man is acknowledged and protected. And so their deaths, when they come, are most likely to be "natural."

Of course natural need not necessarily be peaceful or pleasant. The agent of death could be disease—distemper, canine hepatitis, even rabies, although the latter is uncommon in the area they inhabit. Big Pup, *Canis latrans*, shares the top trophic level with other super-predators and it is possible, although improbable, that he may fall prey to a particularly bold golden eagle or to the cougar who prowls the mountain west of the valley and whose senses are at least as keen as his own. It may happen that he will break a leg during a chase, or be severely injured in a territorial dispute with another coyote, and be so incapacitated that he will be unable to hunt and will starve to death—although if, by then, he has a mate she would certainly help him out. Even without her, the toughness and resiliency of his race could well pull him through.

But in this, his second springtime, as he lofts his jubilant voice above the stillness of the desert, Big Pup has every reason to be content with his lot. He cannot foretell the future, any more than you or I, and unlike us he is not given to pondering such abstractions. But he has a very good chance of enjoying the pleasures of coyotehood for many years to come. And we may reasonably picture him some ten or twelve years from now, old, tired, somewhat arthritic, somewhat shrunken, lying down one afternoon in the shade of a juniper, taking a long look across the valley of his life, then resting his head on his forepaws and letting death come, meeting it with acceptance, with dignity, without fear and without regrets.

4 On the Trail of the Trickster

"WE WERE DOWN by the Mexican border, near Columbus. It's mostly desert scrub around there, but there is this grove of ten cottonwoods. It's a good place to band birds, and it's also used as a resting place by the coyotes in that area. . . ."

Dr. John Hubbard, director of the endangered species program for the New Mexico Department of Game and Fish, was reminiscing with me about coyotes he had met. "On this particular morning we saw this coyote trotting up towards the grove. When he got close he spotted us and stopped short. But he didn't run away. Apparently he'd planned to visit the grove that morning, only we'd beaten him to it. And so all morning long he sat there on his haunches, facing us from 150 feet away, sort of dejectedly yammering at us. Now and then he'd lapse into a coma of self-pity, then start whining, as if to ask us to leave. Finally he slunk off, a picture of frustration and disappointment."

"Didn't you feel kind of mean," I asked, "keeping him away from his grove?"

"Yes, and kind of rude, too. As if he'd had prior reservations to the grove and I didn't have the courtesy to defer to his priority."

To the average man with a gun, this coyote would have been an easy and probably irresistible target. "It's surprising," Dr. Hubbard said to me, "how often you can see a coyote in an absolutely vulnerable situation. I've seen a dozen like that in the last year. And yet they say how wily the coyote is. Could a mutation be producing stupider animals?"

But to me, another remark of Dr. Hubbard's suggested a clue to this coyote's boldness: "I have never shot a coyote, and I never expect to do so," he told me. There is a widely held belief that a coyote can tell at a glance whether a man is armed or not, and will govern himself accordingly. I do not know how much fact and how much imagination go into this claim. But such an ability, if real, would not surprise me too much—it would be akin to the faculty of many herd animals to discriminate between a wolf or a lion on the prowl and one who is merely passing by. And it may have been well within the power of so perceptive and intuitive an animal as the coyote to recognize that Dr. Hubbard meant him no harm.

Certainly, in earliest pioneer days the coyote showed little fear of man. It took time and much sad experience to make him wiser and warier. He seems to have been a much more social animal then than now and was frequently described as moving in packs like the wolf. Excepting at special attractions, such as the carcass of a large animal or a town dump, big gatherings of coyotes are seldom seen today, and established family groups similar to the wolf pack exist only in protected areas such as the national parks. Persecution forced the coyote to adopt more solitary ways, and since he subsists largely on small game that he can catch unassisted, he has been able to do so. This has allowed him to survive in regions where the big gray wolf has been exterminated: a hunter of large game, *Canis lupus* would not or could not abandon the pack organization which made him highly vulnerable to man.

Early reports tell of a hundred or more coyotes being sighted in a single day, of packs of twenty to thirty chasing deer or antelope or a straggling buffalo, of bands of coyotes ringing the campfire at night. Indeed they frequently were bold enough to slip right into camp, and even after decades of not-so-peaceful coexistence with the white man some still would take chances. As late as 1895, Charles F. Lummis wrote in *The Land of Sunshine: A Southwestern Magazine*: "I have known him [the coyote] rather intimately for near a dozen years, and only trust that the pleasure has been mutual. . . .

> . . . He has none of the wolfish ferocity and none of the wolfish seriousness. He is a wag—and like most wags, timid; though I deem it no honor to their intelligence that many call him a coward. Do they expect a thirty-pound wild dog to attack man? But nothing can be more ridiculous than fear of him. A hunter would sit down as unconcerned amid a thousand coyotes as if they were rabbits—unless he had something stealable. . . . I have several times had a coyote step across me while I slept; and among the diversions of our wedding journey in the wilderness was the waking one night to find two coyotes fairly over us, trying to get the saddle-bags from between our heads and the big pine-tree which was our hotel. It needed no more than the creak of an eyelid to send the interlopers flying.

That the first Anglos who invaded the wilderness of the West found this "prairie wolf" so unafraid tells much about the coyote's previous relations with man. The relationship began many thousands of years ago, in an Ice Age world whose history we read not in engraved stone monuments but in a stone projectile point embedded in a mammoth's spine. The Southwest abounds in vestiges of early human occupation or visitation: Gypsum Cave in southern Nevada; Ventana, Lehner, and Naco caves in Arizona; Bat, Folsom, Sandia, and Clovis caves in New Mexico, to name but a few of the more important sites. Whether the Paleo-Indians reached the Southwest 12,000 or 25,000 years ago or even earlier, whether they were preceded on the North American continent by yet more ancient humans who, some experts have speculated, may have crossed the Siberian-Alaskan land bridge during the Sangamon Interglacial period before the Wisconsin glaciation 75,000 years ago and later died out—these are fascinating but highly controversial questions which I will leave to the disputations of archaeologists. The point I wish to make is that during many millennia before the first Europeans reached the Southwest, man and coyote adapted to and learned to appreciate one another.

Of course, just as Dr. Hubbard trespassed on his coyote's cottonwood grove, man himself was originally an interloper in the coyote's North American domain. As far back as the late Pliocene or early Pleistocene—perhaps a million and a half years before the first human mammoth- or bison-hunter appeared on the scene— the coyote was here. His ancient remains have been found scattered over much of the United States. He himself was descended from a proto-coyote who was also the probable progenitor of the Asian and African "coyote," the jackal, or *Canis aureus*. It might surprise you, as it did me when I discussed the subject with Dr. Ronald M. Nowak, a paleoecologist and staff biologist with the Fish and Wildlife Service in Washington, D.C., to learn that the canine family probably originated in North America and migrated to the Old World across the Beringian land bridge as did the horse and the camel. I always had assumed that, like the elephant and man, the wild dogs had traveled in the opposite direction, *from* the Old World *to* the new. For some reason, I also had assumed that the coyote was an offshoot of a primordial wolf. Not so, Dr. Nowak informed me. *Canis latrans* is more primitive in his anatomical structure than *Canis lupus*, and is probably ancestral to him and through him to many breeds of *Canis familiaris*, the domestic dog. And so, if you could trace your pet canine's pedigree a million and a half years or so back into the dim past, you would come to a coyote or coyote-like carnivore who pursued or scavenged the since-defunct elephant, horse and camel on what are now the prairies of the Dakotas or deserts of Arizona but were, in those "high and far-off times," neither prairie nor desert.

With dire wolves and sabertooth tigers still roaming his world, it seems unlikely that North America's Ice Age hunter would have been very impressed on first meeting the coyote, and that he would have singled him out as the hero of myth and legend that he was destined to become to the tribes of the Southwest. But then we do not know, although we may conjecture what we will, that these Paleo-Indians looked upon *any* animal as something more than a menace to their lives or a resource to be exploited. We do know that they exploited the mammoth, mastodon, camel, horse, great bison, and giant sloth. And some authorities have theorized that they did so with an efficiency and with consequences that presaged (but for an eleventh hour change of heart) our own exploitation of the modern bison, and that today's Russian and Japanese whale hunters bid fair to emulate.

Whether or not he was responsible for the final extinction of the great beasts, whose populations were already decimated by climatic change, Ice Age man in North America certainly killed more than he needed. A common practice, for example, was the stampeding of a whole herd of bison over a cliff—piles of bones remain as evidence. But how did the Paleo-Indian feel about the animals he hunted? Did he have, even embryonically, the notion of spiritual kinship with his prey so characteristic of his descendants several millennia later? Perhaps he did, but of that he left no evidence. He was roughly contemporaneous with the Cro-Magnon artists who painted the glorious animal frescos on the cave walls and ceilings at Lascaux and Altamira. But there are no Lascaux or Altamiras in America.

The great Ice Age mammals were all dead by the time unknown Indian artists began scratching and painting pictures of deer and bighorn sheep, lean coyotes with lolling tongues, eerie divinities and, to us, indecipherable symbols on the walls of hidden canyons throughout the Southwest. The sandstone cliffs had long ceased echoing the trumpeting of the last mammoths by the time Desert Culture hunters-and-gatherers twisted split twigs into figurines of deer, bighorns, and pronghorn antelope and deposited them in remote caves within the Grand Canyon.

On one of my river trips I climbed up into one of these caves. In a dark, low-ceilinged side-chamber into which I crawled, guided by the beam of my flashlight, I knelt sifting the sand and crumbled rock of the cave's floor through my fingers. In a couple of minutes I found what I was looking for: a beautifully fashioned little figure of a desert sheep, complete with spiraling horns. I reburied it where it had lain for 4,000 years, but not until I had studied and fondled it for a while. What did it mean, I wondered? What did it mean to the man who made it? What did it tell me about that man?

It meant a great deal, I thought. In view of its inaccessibility, this cave could not have been a dwelling—researchers have found no sign of habitation there. So it must have been a shrine, at which this ancient Indian sought to propitiate higher powers that they might aid him in the hunt. I could picture that naked or near-naked hunter scrambling down the cliffs and taluses to the river, breaking off twigs from the willows that bordered the turbid, gurgling water, bending and twisting and braiding them into a simulacrum of the animal he hoped to bring back to his family campfire up on the plateau back of the canyon rim. Then I envisioned him climbing up to the cave, as I had done, and depositing his offering in the cave's inner sanctum, with a prayer to the guardian spirits of the hunt. This man could conceive an abstraction, bighorn sheep, and give the abstraction a symbolic physical representation, the figurine, and infuse the symbol with numinosity. He must have had a vision of another dimension to his life and the life of the world about him, a spiritual dimension that transcended and transfigured the realities apparent to his senses. Perhaps he prayed, as Indian hunters not too corrupted by the white man's jejune pragmatism still pray four millennia later, to the spirit of the sheep, immanent in the plaited twigs: "Forgive me, brother sheep, for I mean to kill you. But if you will heed my prayer and come to me, my brother, I will honor you in death as I respect you in life."

"You whites assumed we were savages. You didn't understand our prayers." Tatanga Mani or Walking Buffalo, a Stoney Indian born in 1871 near Morley, Alberta, was addressing an audience in London at the age of 87. "You didn't *try* to understand," he went on. "We saw the Great Spirit's work in almost everything: sun, moon, trees, wind, and mountains. Sometimes we approached him through these things. Was that so bad?"

We can dismiss the Indian's spiritual view of nature as a primitive animism, and his creation myths with their protean gods in animal garb as grotesque superstitions. Or we can regard him as a genuine pantheist, and see in his sagas of the emergence of man through a succession of subterranean worlds of pre-human flux poetic parables of man's rise to consciousness, rich in Jungian overtones.

We can be cynical if we will and speculate that had the Indian possessed a technology comparable to the European's he would have brutalized his environment just as harshly as we have done. Or we can view him as the quintessential conservationist. I try to avoid an overromanticized picture of the aboriginal Indian, but I confess that I find in him much more to admire than to scorn. In his quoted utterances, he revealed a mystical understanding of the oneness of all life, a reverence for the land which he shared with other living things, that made him a soul-brother to the likes of St. Francis and Albert Schweitzer. In the long agony of his defeat and dispossession by the white man, the Indian rose time and again to speak of the land he had loved and lost with a sincerity, an eloquence, a poignancy that tug at the heart, and his comments on the white man's perception and treatment of nature are no less devastating for being understated. Listen to some of these voices, orchestrated by T. C. McLuhan in her anthology, *Touch the Earth: a Self-Portrait of Indian Existence*:

"Everything as it moves, now and then, here and there, makes stops," an old Dakota wise man said in the late 1800s. "The bird as it flies stops in one place to make its nest, and in another to rest in its flight. A man when he goes forth stops when he wills. So the god has stopped. The sun, which is so bright and beautiful, is one place where he has stopped. The moon, the stars, the winds, he has been with. The trees, the animals, are all where he has stopped, and the Indian thinks of these places and sends his prayers there to reach the place where the god has stopped and win help and a blessing."

In 1796, an Indian chief told the governor of Pennsylvania: "We love quiet; we suffer the mouse to play; when the woods are rustled by the wind, we fear not."

Contrast these attitudes with the white man's fear of raw nature, as diagnosed by Oglala Sioux Chief Luther Standing Bear in his autobiography published in 1933:

> The white man does not understand the Indian for the rea-
> son that he does not understand America. He is too far
> removed from its formative processes. The roots of the tree of
> his life have not yet grasped the rock and soil. The white man
> is still troubled with primitive fears; he still has in his con-
> sciousness the perils of this frontier continent, some of its

vastnesses not yet having yielded to his questing footsteps and inquiring eyes. He shudders still with the memory of the loss of his forefathers upon its scorching deserts and forbidding mountain-tops. The man from Europe is still a foreigner and an alien. . . .

We did not think of the great open plains, the beautiful rolling hills, and winding streams with tangled growth, as "wild." Only to the white man was nature a "wilderness" and only to him was the land "infested" with "wild" animals and "savage" people. To us it was tame. Earth was beautiful and we were surrounded with the blessings of the Great Mystery. Not until the hairy man from the east came and with brutal frenzy heaped injustices upon us and the families we loved was it "wild" for us. When the very animals of the forest began fleeing from his approach, then it was that for us the "Wild West" began.

The "hairy man from the East's" fear and hatred of those aspects of nature he could not dominate or control—the winds, the droughts, the locusts, the wolves—and his drive to subjugate the earth and to wrench wealth from her very bowels, these were as foreign to the Indian's soul as they were impossible to his technology. "The white people never cared for land or deer or bear," an old Wintu woman of California cried as she saw the land in which she lived devastated by hydraulic gold mining. "When we Indians kill meat," she said,

we eat it all up. When we dig roots we make little holes. . . . When we burn grass for grasshoppers, we don't ruin things. We shake down acorns and pinenuts. We don't chop down the trees. We only use dead wood. But the White people plow up the ground, pull up the trees, kill everything. The tree says, "Don't. I am sore. Don't hurt me." But they chop it down and cut it up. . . . They blast rocks and scatter them on the ground. The rock says, "Don't. You are hurting me." But the White people pay no attention. . . . How can the spirit of the earth like the White man? Everywhere the White man has touched it, it is sore.

In 1909, Chief Plenty-Coups of the Crow Nation gave a farewell address at the Little Bighorn council grounds in Montana. "A few more passing suns will see us here no more," he foretold, "and our dust and bones will mingle with these same prairies. I see as in a vision the dying spark of our council fires, the ashes cold and white. I see no longer the curling smoke rising from our lodge poles. I hear no longer the songs of the women as they prepare the meal. The antelope have gone; the buffalo wallows are empty. Only the wail of the coyote is heard. . . ."

How poetic, pathetic—and perhaps prophetic! "Only the wail of the coyote is heard. . . ." The other animals would join the Red man in oblivion—but the coyote would survive. One of the "first animals" in the cosmogonies of many Western tribes, how fitting that he should also be there at the end! Coyote the resourceful, the tough, the enduring, would enjoy the last taunting howl when all the rest is silence. When I came across Chief Plenty-Coups' words in McLuhan's anthology, I was reminded of all the white sheep-men and trappers and hunters who had said to me or written to me, "The coyote will be here long after we are all gone!" Some said it with bitterness, some with resignation, but all in a tone of at least grudging admiration.

No other animal so captured the imagination of the Indians of the West as did the coyote. He is the hero or anti-hero of countless myths, folktales, and legends, in which he appears as a spirit being, or a humanized coyote who usually comes to no good, or as the real animal the tribesmen see trotting past the mesa.

"Throughout Navaho mythology, Coyote (Mah-ih) is a dynamic figure," Ruth DeEtte Simpson writes in an article in *The Master-key*, "The Coyote in Southwestern Indian Tradition" (March-April 1958):

> In the Navaho coyote myth, it is told how he brought his people ashore from the original island; how he got his power from the rainbow; how his people, the Yellow Corn People, followed him, heard him sing; how he became a Yei. Even after that, the rainbow guided the people in their travels. Later, a young relative of Coyote's visited the houses of the ancestral Coyote People and brought back the ceremonies of the Navaho and also brought them corn. All these the youth gave the Navaho through his brother and then returned to live with Coyote People.
>
> When the Navaho come to this world, Coyote steals the children of the Water Monster and thus precipitates the advance of Man and all other creatures to this world.

Among the Maricopa, it is believed that the Creator left the Maricopa before they had learned how and when to plant their seeds. They asked him to return and tell them and he promised to come. While they waited, all the Maricopa went to sleep and only Coyote heard the instructions, so he taught the people when and how to plant.

In the myths of the Zuñi, who have a particular trust in and fondness for Coyote, it was Coyote-Being who taught man how to be a good hunter, and gave him the proper rites and songs. The Paiute, on the other hand, believe in a sort of Garden of Eden in which all animals played together, spoke the same language, and ate grass. Only Coyote was not satisfied, and instead of attending a council of all the animals he curled up and went to sleep. While he slept, the Evil one gave him meat, which he liked very much, and when the other animals began to straggle back from the council he started killing his rabbit friends and others. Soon the animals were afraid of Coyote and suspicious of one another, the common language was lost, and now no one loves or trusts Coyote.

The Miwoks of California relate that when Coyote had finished working on the world and had made the other creatures, he called them into council to discuss the creation of Man. Each animal had his own ideas, each ridiculed the other's. The cougar said Man should have a mighty voice and sharp, grasping claws. Pooh!, the grizzly said; Man should have prodigious strength, and be able to move silently but swiftly. What Man needed, the buck said, was a magnificent set of antlers. No, not antlers, objected the mountain ram—they would just get caught in the thickets. Better that he have two heavy spiralling horns. The owl would give Man wings. That would be dangerous, demurred the mole; better he be able to burrow into the cool, soft earth. Lastly, the little mouse squeaked that Man should have whiskers so that he could feel his way in the dark.

When all had had their say, it came Coyote's turn to speak. As J. Frank Dobie tells the tale, Coyote

declared that he could hardly keep awake listening to such a pack of egotistic nonsense. All that any speaker wanted was to make Man a mere imitation of himself. If this was the idea, why not pick out any cub of the lot and just call it Man? And nobody had even suggested a mind. As for himself, he knew he was not the best-formed animal that could be made. It was

well enough to have a voice. The buffalo bull had the strongest in the land, and he was utterly stupid. The shape of the
grizzly bear's legs and feet, which enabled him to stand erect
and reach out, was certainly desirable for Man. The grizzly
was also happy in having no tail, for he had learned from his
own experience that that organ was only a harbor for fleas.
The buck's eyes and ears were pretty good, perhaps better
than his own. Then there was the fish to think of. He was
naked and could keep cool in the summer and be free from
fleas the year round. Man should be hairless, except for a
patch to protect his brains. His claws ought to be long and
flexible like the eagle's, so that he could grasp things. Yet,
after all the separate gifts were added together, the animals
must acknowledge that nobody but himself had the necessary
wit to supply Man. He would be obliged, therefore, to make
Man like himself in cunning and adaptability.

And so it was done. Each animal took a lump of clay and made a
model of Man as he envisioned him, then went to sleep. But
during the night Coyote destroyed all the models by urinating on
them. Then he fashioned his own, just as he had described it, and
breathed life into it. And when the animals woke up at dawn, there
before them stood Man.

One day in Window Rock, the Navajo tribal capital, I paid a call
on Carl Nelson Gorman, nationally known artist and a former
instructor in Navajo culture at the University of California at
Davis. "In our traditions," Gorman explained to me, "different
animals serve as different symbols. The Crow is a symbol of vanity;
the Owl is a symbol of death—an owl hooting near a hogan is a
warning that death is near. And the Coyote is a symbolic representation of man, with his dual potential for good and for evil."

Even at his best, the Coyote in Southwestern Indian lore distin-
guishes himself as a tricky fellow. So long as he associates with the
Holy People, and uses his cunning to help others, he suffers no
retribution for his tricks. But when he uses it to try to outwit his
fellows and to satisfy his own greed, lust, or idle curiosity, he
almost always succeeds in out-foxing (out-coyoting) himself. The
Southwestern tribes have an immense body of folk tales in which
Coyote appears as the archetypal Trickster tricked—the counterpart of the Reynard the Fox stories of Europe, although innocent of
the bitterness and satire of the latter. A surprising element of
Coyote's character in these tales, in view of the real animal's

acknowledged wiliness, is his unshakable gullibility, his willingness to believe anything he is told.

By and large, Coyote serves in Navajo tales as a model of what
not to be, what not to do. " 'Trotting Coyote' is a representative of
socially unacceptable behavior," Robert A. Roessel, Jr., and Dillon
Platero write in *Coyote Stories of the Navajo People*. "His ultimate
misfortunes are legendary proof of the disastrous effects of anti-
social conduct. The ultimate victory and good fortune of those
whom 'Trotting Coyote' tries to trick, cheat or destroy affirms the
eventual triumph of justice and morality. The stories thus
strengthen and reinforce moral values, social harmony, and cultural norms by endowing them with the prestige and power of
antiquity, as well as with the sanction and affirmation of the
supernatural."

Of Roessel's and Platero's compilation of fourteen coyote tales,
"Coyote and the Lizards" is typical:

Coyote was always happiest when he was spying on someone
or prying into his business. One day, when he saw a group of
lizards playing a game that was strange to him, he trotted over
to learn all about it.

The lizards were gathered on top of a big, flat rock with one
sloping side. They were taking turns sliding down that steep
slope on small flat rocks.

Each time, after his slide, the slider picked up his rock at
the bottom of the slide and carried it up the hill on his back.

Coyote trotted over to the rock and sat down nearby. The
lizards pretended not to see him. They went on with their
play, as if he were not there at all.

Coyote didn't like that. He wanted to be noticed at least.
He moved a little closer and began talking to the lizards.

"You seem to be having a lot of fun," he said. "What do you
call your game?"

"We just call it sliding," one said.

"Sliding, eh?" Coyote was trying hard to be friendly. "It
looks so interesting I'd like to join you."

All the lizards turned and looked at him coldly.

"You are not a lizard," one of them said. "Go play your own
games. You don't know ours."

"But I can learn," Coyote insisted. "Really, it looks very
simple. I'd just stand on the rock and slide down. Let me try
it. Just once."

"This game is very dangerous. You'd get killed," an old lizard told him. "The first time you would be all right, but the second time, when you ride the big rock, you'd be smashed flat."

Coyote didn't believe a word of that. None of the lizards had been smashed, so why should he? He kept begging them to let him try it, just once.

"Well, just once, Cousin," said the oldest lizard after hearing Coyote begging. "You can ride the small flat rock, but don't ask to ride the big one."

Of course Coyote intended to ride the big one, also, but he didn't say anything about that at the time. He decided to try the little one, and show them how well he could do it. Then he would persuade them to allow him to try the big one.

The lizards looked sour as they placed the small flat rock in position for him.

"I don't know why you want to play our games," one of them said. "I happen to know you have lots of games to play. I've seen you chasing cottontails and kangaroo rats and all sorts of creatures. I should think running races would be more to your liking. You *are* a fast runner."

Coyote didn't answer. He stepped out on the flat rock. It tilted down onto the runway and—Zi-i-ip!—away he went like a streak of lightning.

Before he reached the bottom of the slide, Coyote jumped off. He picked up the rock and trudged back up the hill with it.

"You see," he panted. "I can do it. Let me use the big rock. Just once."

The lizards looked at him sternly.

The oldest one said, "We warned you. We didn't want you to try the big rock, but your life is your own. If you want to risk it in this way, it is your own fault if you get smashed flat."

The old lizard told the young lizards to get the big rock for Coyote. They moved away silently and came back with it. They placed the big rock on the edge of the runway. Then they stood back.

Coyote was not at all afraid. He ran out onto the rock, tipped it a little, and once more he was sliding very rapidly down the runway. But the big rock caught on a smaller one half way down the slide. The rock flipped into the air, taking Coyote with it.

Coyote was frightened half out of his skin. His ears were

flopping and his paws were clawing the air. He wasn't at all proud of himself, as he had been on the first ride.

He hit the ground and rolled over. He saw the big rock coming down on top of him.

"I should have listened," he thought. "I'm going to be smashed flat, just as they said."

The big rock fell and smashed Coyote.

The lizards stood looking down at him.

"Poor foolish Coyote," the oldest lizard said. "He's no friend of mine, but still it makes me sad to see him smashed so flat."

"And right in the middle of our runway," said one of the young lizards.

"It wouldn't be right to leave him there. But he's going to be very heavy for us to move," said another.

"It would be simpler to bring him back to life," said a third lizard. "Then he could leave without us having to move him."

"You have a very good idea," said the oldest lizard. "Come on, boys."

Single file they slid down to Coyote and made a tight circle around him so they could work their magic in private. In their own secret manner they brought him back to life.

"Now go on your way, Coyote," the oldest lizard told him. "And, after this, don't try to play lizard games. We don't want this to happen to you again."

Coyote was glad to be alive. He got up and dashed for home as fast as he could run.

Perhaps as remarkable as Coyote's gullibility is his indestructibility—or, more accurately, his powers of resurrection. In story after story he is burned, drowned, starved, smothered, crushed, or dashed to death from a height, and he keeps coming back—and none the wiser for his misfortunes. His end is not always tragic, however; sometimes it is just funny. "The Bluebird and Coyote" tale, quoted by Frank Russell in *The Pima Indians*, tells how Coyote got his dust-colored coat:

The bluebird was once a very ugly color. But there was a lake where no river flowed in or out, and the bird bathed in this four times every morning for four mornings. Every morning it sang:

There's a blue water, it lies there.
I went in,
I am all blue.

On the fourth morning it shed all its feathers and came out in its bare skin, but on the fifth morning it came out with blue feathers.

All this while Coyote had been watching the bird; he wanted to jump in and get it, but was afraid of the water. But on that morning he said, "How is this all your ugly color has come out of you, and now you are all blue and gay and beautiful? You are more beautiful than anything that flies in the air. I want to be blue, too." Coyote was at that time a bright green. "I only went in four times," said the bird; and it taught Coyote the song, and he went in four times, and the fifth time he came out as blue as the little bird.

That made him feel very proud, because he turned into a

blue coyote. He was so proud that as he walked along he looked about on every side to see if anyone was noticing how fine and blue he was. He looked to see if his shadow was blue, too, and so he was not watching the road, and presently he ran into a stump so hard that it threw him down in the dirt and he became dust-colored all over. And to this day all coyotes are the color of dirt.

Some of the stories can be delightfully picturesque, such as one told by Ruth Simpson:

It will be difficult to look at the heavens at night without remembering the Hopi explanation for the stars being scattered across the sky: Coyote was given a big jar to carry when the people came to this world. He was told not to lift the lid; the jar was heavy and Coyote was curious to know what he was carrying. He set the jar down, lifted the lid and peeked inside. All the stars rushed out and they burned his nose as they flew past. Poor Coyote could only catch a few and these are now in their proper places in the skies (the constellations), the others are just strewn about, and some of them, being insecurely fastened in place, fall back to earth. This is why there are few constellations, why there are shooting stars and why Coyote has a black nose.

These tales reveal an amused tolerance and often a genuine fondness for the coyote. That he was usually portrayed as a trickster is understandable considering the nature of the animal himself: the coyote hunts more by wit than by brawn, and tales of his ingenuity and craftiness in securing his food, told not by Indians but by "objective" Anglo observers, are legion. That they should have found him amusing is also natural. Unlike the potentially menacing bear or the stately wolf, the coyote does not require that he be taken too seriously. His is a mirthful figure; coyote pups are as beguiling in their play as pups of the domestic dog, and even the adults can be hilarious at times. Describing how coyotes capture ground squirrels, Hope Ryden writes: "A sudden low scurry either secures the prize or sends a hunter into a nose dive. On several occasions I have seen coyotes somersault, head over tail, while making unsuccessful lunges at ground squirrels."

Mexican popular speech often reflects Indian attitudes, and contains a good number of *indianismos*. The word *coyote* itself derives from *coyotl* in Nahuatl, the language of the Aztecs;

Hispanicized to *coyote* (co-yo'-tay), by 1830 the word was being adopted by English speakers in lieu of the term "prairie wolf." (Today it is pronounced either "ky-oh'-tee" or "ky'-oat"—take your pick.)

In the Mexican vernacular, according to Dobie, *coyote* means: "a pettifogger, a thief, any kind of shyster or go-between, a curb-stone broker, a fixer who has 'pull' to sell, an oil or mining scout with 'practical experience' in selling leases, also the respectable Minister of Mines, a drink of mixed beer and brandy. As [writer Carl] Lumholtz puts it, 'The regard that the Indians have for their Mexican masters is shown in the name by which they refer to them —coyotes.' . . . Mexicans call a 'wolf' among women *un coyote*; they call bastard children *coyotitos*. Without aspersion they call also the last child in a family a *coyotito*." But not all connotations are disparaging. A shrewd man can be *muy coyote* without being a crook.

In part, the Indians' traditionally benign attitude towards the coyote may have derived from, or be reflected in, their belief in his curative powers. In the Pima-Maricopa creation myth, after the Creator died and while he was being cremated, one of his children, Coyote, stole his heart and ate it. Soon Coyote became ill, and illness followed illness until he had contracted every disease of man. Then he ate grass and became well, although terribly thin. Having suffered all diseases and recovered, Coyote could cure them all and became the only animal with that power.

But, according to Ruth Simpson, there may have been some basis of fact in the widespread Indian belief in the coyote's curative power and in his benevolence:

> In at least some degree [it] may spring from the fact that coyotes do frequently stay with a person whom they find sick or injured. The coyote will stand guard against enemies or animals threatening the victim, and will lie by him, licking the wound. [Dobie gives an instance of this.] Many times, coyotes have come to men (Indians and white) and led them to some injured person or lost child. Most important is the fact that the presence of the coyote gives the lonely man in trouble new courage, new heart to fight for his life until help comes. As late as 1839, the Kiowa held a public thanksgiving for a coyote who had saved in this manner one of their hunters wounded far from camp.

Since coyote skins are used in traditional ceremonial costumes of certain tribes, it seems likely that a few animals were killed for this purpose throughout the centuries. Occasionally, perhaps in times of famine, coyotes were hunted for food. Dr. Douglas Schwartz, director of the School of American Research in Santa Fe, told me that excavations of a fourteenth century pueblo just south of Santa Fe had turned up bones of four coyotes who had been killed by the inhabitants and presumably eaten. But these four coyotes represented only .3 percent of the 1500 individual animals whose remains were found in the ruins. "There were undoubtedly lots of coyotes around," Dr. Schwartz told me, "and some were eaten, but they were not utilized as a normal part of the food resource. Probably they were too elusive—or too stringy." (The idea of coyote stew does not tempt my French palate, although I received a letter from a young man in Montana, who styles himself "half bum, half woodsman, and half casual laborer," and who writes: "I would emphasize that coyote meat is very good eating—to my taste better than venison—but it seems necessary to simmer it gently for an hour or two to make it tender. It has a peculiar flavor that I have never tasted in any other kind of meat.")

But for centuries before the arrival of the white man and his domestic stock, and particularly before the arrival of the Anglos in the Southwest in the nineteenth century, man and coyote lived together in peace. The coyote had little to fear from man, and man had no reason to fear the coyote. Occasionally a Navajo might suspect a coyote crossing his path of being a "human wolf" and would hasten back to his hogan to flee this evil reincarnation of a departed relative, but the coyote posed no threat to the settled agriculturalist tribes. And though to hunting tribes he might be a direct competitor for game, Indian hunters never considered, as Anglo hunters do today, that game was put on this earth exclusively for their use. The deer was his own master, and so was the rabbit. Man might make use of the game resource, after observing the proper rites, but he no more claimed property rights to game animals than he did to the land itself. Both he was willing to share, and the coyote and the eagle had as much right to hunt as he.

What has a century of Anglo domination done to these traditional attitudes and beliefs? To what extent have white schoolteachers, white missionaries, white traders, white Bureau of Indian Affairs administrators, succeeded in inoculating their Indian charges with the Western Christian ethic, which claims Divine sanction for man's self-arrogated dominion over the natural world and the right to tame, cow, and mutilate it to suit his wants? How much has the Anglo stockman's congenital predator-hatred rubbed off on the traditionally tolerant Indians—particularly those, such as the cattle-raising Papago or the sheep-herding Navajo, who have acquired domestic stock of their own?

I cannot presume, on the basis of a few days spent on the Navajo and Hopi reservations, and a few interviews with Indians and with whites knowledgeable on Indian affairs, to give an authoritative answer to these questions. I can only give a few impressions, and quote a few informants.

One sentence, in one report, offered a clue to current Indian attitudes towards the coyote. *Perspectives on Land Use: Problems of Wildlife Management Viewed from Varying Cultural and Socio-Economic Perspectives*, by Drs. Gary R. Olsen and Christian J. Buys, reporting on a series of forums sponsored by the New Mexico Humanities Council, described a prevailing attitude that "human needs, real or imagined, rightly take precedence over those of other creatures." And it commented: "Even the American Indians who participated (and from whom we hoped for a less man-centered view) ultimately had recourse to what might be called the Anglo orthodoxy as regards predators."

Was this really so? Bill Bishop, a young, active environmentalist who has taught among the Navajo, replied to my question: "I've asked quite a few Navajos how they feel about the coyote. One old sheepherder, who didn't speak much English, spat on the ground and said, 'Coyote shit God damn!' I guess he didn't have much use for them. One old lady I talked to outside of her hogan said she never was bothered by the coyotes. She always went out with her flock and watched it while the coyotes watched her. But one rich Navajo, who owns a large number of sheep and hires a man to look after them, told me he'd ordered his herder to shoot every coyote he saw.

"You'll probably find," Bill went on, "that the 1976 attitude on the reservation is just as varied and not very different from what it is off-reservation. Probably the more traditional a Navajo is, the less he worries about coyote depredation. He tends his flock, he respects the coyote, and perhaps he fears him a bit. There's still a lot of talk among Navajos about witches, shape changers, 'wolf men.' I had a very good student who missed class for a week. When he came back the next week he said he was sorry but he'd seen a wolf man and was scared to leave home."

"The coyote tales," Bill answered another of my questions, "seem to be dying out. I once asked a group of my students if they knew many of them. Most of them had never even heard of them!

A few knew there were such stories, but they'd never heard any themselves. Only two or three actually remembered having heard them told in their homes."

According to Dr. Robert W. Young of the University of New Mexico who is compiling a Navajo-English dictionary, more and more young Navajos grow up without hearing the coyote stories. "There're really two categories of people on the reservation," he said: "The older and the younger. The younger are pretty much acculturated; they've revised their thinking about the coyote and tend to view him as a threat to the livestock. Coyotes do kill lambs, and many Navajos today will shoot them or poison them for it. I doubt that thirty or forty years ago very many Navajos destroyed coyotes. When I was editor of the *Navajo Times*, in Window Rock, I had Navajos come to me more than once and ask me to kill coyotes for them. They were losing sheep and wanted the coyotes killed, but they wouldn't do it themselves.

"But even today," Dr. Young continued, "Navajos don't believe in destroying anything wantonly. Everything, as they see it, has its place in the scheme of things, and that includes the coyote."

When I related to him Bill Bishop's story about the student who'd seen a "wolf man" and been afraid to come to class, Dr. Young nodded. "In actual life," he said, "the coyote plays an odd role. He is a scavenger, and sometimes he digs into graves. And thus he gets contaminated. Navajos believe that men are a compound of good and evil. Burial after death separates the two. Ghosts are evil and will often take the shape of a coyote. If a Navajo is going along a road and a coyote crosses in front of him he won't proceed: he sees it as an ill omen."

The Papagos, too, believe that coyotes can be harmful, although in a different way. As Dr. Bernard Fontana of the University of Arizona explained, "Papagos recognize a 'coyote sickness' that people can get if they have transgressed the dignity of the animal. A whole series of living things are considered to have power. Papagos talk of a 'coyote way,' just as we speak of the 'American way.' If you transgress the dignity of an animal, eventually you get sick; then you have to go to a medicine man who diagnoses which 'way' you have transgressed and says the right songs.

"Stepping on a coyote's track can be a transgression of his 'coyoteness' and can cause coyote sickness. You can get coyote sickness by killing a coyote or merely by eating a melon that a coyote has bitten into.

"Aboriginally," Dr. Fontana went on, "the Papago tribe had a dual division into moieties: the buzzard moiety and the coyote moiety. For those who belonged to the coyote moiety to do any harm to a coyote was absolutely taboo. But Papagos as a whole never had any hostility to the coyote: it was live and let live. Nowadays, some may have adopted the white cattlemen's outlook. But even now most Papagos have a positive attitude towards the coyote."

A Bureau of Indian Affairs (BIA) range-management specialist at Sells disagreed. "Some of the older people still cling to the ancient superstitions," he said, "and hold the coyote in some degree of reverence. But most middle-aged or younger Papagos are indifferent to him one way or another. Some would eliminate it totally." I inferred, perhaps erroneously, from the tone of the specialist's talk that his own regard for the coyote was no greater than his regard for the Papagos' "superstitions," but he admitted that drought and overstocking were more serious problems on the reservation than coyote depredations on calves, although these do occur. He also conceded that there was only very limited shooting of coyotes by Papagos, no trapping, and no use of poisons whatsoever. Apparently Papago coyote-haters are in a small minority.

The Hopi, also, seem to be relaxed about the coyote—which is not surprising since they are principally agriculturalists. According to Marvin Jones at the BIA Hopi Agency at Keams Canyon, the Hopi only run some 2,000 sheep and less than 6,000 cattle. Lamb and calf losses were high enough to warrant the hiring of one trapper in 1975, Jones told me, but he caught eight dogs for every coyote he took. The worst damage, he said, was caused by coyote-dog hybrids.

The Hopi, like most Indian tribes, are divided into clans. Each clan has its own traditions and its own rites; a Hopi's primary loyalty is to his clan; it is the heart of Hopi society. There are twelve principal clans, of which one is the Coyote Clan. I spent an hour one morning talking with an old lady of the Coyote Clan, in her small stone house at the foot of a mesa. She happens to be a world-famous potter but that was not why I sought her out: I was interested in her clan affiliation and whatever special insights she might have on the coyote. But she asked that I not mention her by name, so I will just call her Mrs. Smith.

She began talking in a quiet, firm voice, her eyes looking down at her skilled hands folded in her lap. "The coyote," she began, "is a very important animal. He is smart. He is enduring. He is unassuming. He is the humblest of all. And yet he wants for nothing." I found it difficult to decide, as she spoke, whether she was talking about the mythical Coyote-Being or about the little wild dog she might have seen trotting down the wash. Perhaps she meant both. The coyote, she said, "possesses every quality of man—both good and bad." Because the coyote is "at the base of the lowest," she said a little later, he has great power. In the Hopi view, humility is the greatest virtue. A traditional Hopi will not *seek* leadership, but there is power in humility. And so the Coyote Clan, although regarded as the "lowest" of the clans, is also viewed as possessing secrets that will ensure the preservation of the Hopi tribe and way of life. And these secrets would be needed. Because unfortunately, she said, the mystical identification with nature, which led her grandfather to talk to each growing plant of corn and "caress it with his hands as he would caress a baby," was rapidly being lost by the young.

I watched the young that night, when I attended a "bean dance" at the First Mesa village of Sichomovi. They seemed as rapt as I, although they had seen the kachinas many times before. The few lights in the central square hardly dimmed the brilliance of the stars, but gave a soft glow to the long façades of low adobe-and-stone houses that rimmed the court. From the two kivas, the underground ceremonial chambers that occupied the center of the plaza, issued the insistent throb of drums, the rhythmical susurration of rattles and tinkling of the bells tied to the dancer's ankles, and the hypnotic, repetitive chanting. During a break in the dancing, I climbed down the ladder into one of the kivas and found a place to stand by the rear wall. A shout and loud stamping resounded from above—a new group of dancers asking permission to enter—and the chief of the kiva shouted back his welcome. The masked spirits, arrayed in white kilts, bright-colored sashes and glistening body paint, quickly backed down the ladder, formed a U-shaped line, and began to stamp and gyrate to the beat of the drum. As the night wore on into the small hours of the morning, groups of dancers succeeded one another, each garbed in a different costume, each wearing a different grotesquely beautiful mask. One group's costume was decorated with gray fox and coyote skins. How much more appropriate, I thought, if animals must be killed for their skins, that they should be used in these celebrations of the great cycles of nature than that they should be sacrificed to someone's sartorial vanity! There were youthful Hopi in the audience—from toddlers to young adults—and they appeared to be as entranced as I was.

"It's just a spectacle to them," Homer Cooyama said to me the next day, as we sat in his house near Oraibi. "They've lost all sense of the meaning of the spirits." Homer, like Mrs. "Smith," is an elder member of the Coyote Clan. But he was not interested in talking about coyotes. He wanted to talk about the future, or lack thereof, of the Hopi way of life, the unconstitutionality and unrepresentativeness, as he saw it, of a Tribal Council dominated by progressives, the capitulation to white ways and white interests. "Go see what they're doing on Black Mesa," he said, referring to the Peabody Coal Company's mammoth strip-mining operation. (I did, a few days later, and knew I'd witnessed rape.) "The young," he concluded bitterly, "they just want to be white."

I had an appointment the next morning, arranged through my friend Harry King, an anthropologist from Santa Fe, to interview two young Indians some thirty miles south of Oraibi. These were not Hopi—they were teen-aged Navajo sisters who spoke good English and whose grandfather, Harry told me, was a well-known "singer" or medicine man. I started down the dirt road that generally follows Oraibi Wash to the Navajo settlement of Leupp on the Little Colorado, and had gone two or three miles when a coyote loped across the road just ahead of me. I stopped the Land Rover, and watched him as he ran a couple of hundred yards across the desert, then stopped and turned to look at me. Perhaps I ought to turn around and go back, I thought; according to Dr. Young that's what a good Navajo would do. But I decided this was just a coyote, not a "wolf man," and so proceeded on my way. I found the girls' compound according to Harry's directions: a small wooden house, a hogan, a ramada, and a corral enclosing a small flock of sheep and goats. Only one of the girls was at home—her sister was seeing a doctor in Winslow—and the old medicine man, it turned out, had died the month before.

But Louva, an attractive fifteen-year-old, was puttering around the yard. So I followed her as she did her chores, and explained that I was interested in her experiences and attitudes with regard to the coyote. "The coyote?" she seemed surprised. "Oh, we've got lots. They're all over the place, especially over there in the forest," she said, pointing east across the road to a wide expanse of sand dunes topped by scrubby trees. "Do you have much trouble with them?" I asked. "Trouble, yes, all the time. First the Hopi kill all our cattle, now the coyote kill all our sheep." How many sheep had she lost? "Oh, we used to have two-three hundred, now we just have those." She pointed to the corral. Did she know any coyote stories? "Oh, yes, grandfather he tell us many coyote stories. I like the coyote stories, but I don't like the coyote! Are you going to kill some coyotes for us?" she asked. "No, I'm sorry," I answered, "I'm just trying to learn about them, not kill them." She seemed disappointed.

At that moment Louva's brother came cantering up on horseback, dismounted and tied his horse to the corral gate. A little later I noticed that the horse had pulled open the corral gate and the sheep and goats were ambling off single-file into the desert. I called Louva's attention to this, thinking that they seemed easy prey for the coyotes, but all she said was, "Oh, they go to eat," and turned back to a comic book she was now reading. Realizing the interview was over, I thanked her and drove off, remembering Bishop's words: "The more traditional a Navajo is, the less he worries about coyote depredation. He tends his flock."

Three days later I was in Window Rock, talking with Carl Gorman. After we had discussed the mythological aspects of Coyote, I asked him how Navajos felt about coyotes raiding their flocks. "We don't like to kill animals," he said. "We feel animals were put here for a reason. However, we feel if a coyote has killed a lamb we have a right to kill him. But we're against the use of poison. Poison kills everything. It's a white man's method."

I paid a call on Roger C. Davis, a tall, good-looking Navajo and director of the Navajo Tribe's Resources Division. Dressed in a well-cut Western suit and boots, Davis looked like a well-to-do rancher, which he is. "Since the ban on 1080 and other toxicants in 1972," he told me, "we have been overrun with predators. Mainly coyotes and wild dogs, occasionally bear. Sheep losses have been climbing every year since 1973. Maybe 1080 was misused as charged by the EPA [Environmental Protection Agency] but it was the only effective tool we had here on the reservation. Last year we joined with representatives of the National Woolgrowers, Cattlegrowers, and Turkey and Poultry Growers and had an audience with President Ford in which we urged him to rescind or amend the executive order banning poisons. But nothing's happened. It looks like coyote predation's just something we'll have to live with. Either we must accept being managed by the coyote or we'll have to go out of business altogether."

When I told Davis that I had inquired at Rocky Ridge, at Ganado, and at Lukachukai if coyote damage had been heavy, and had been told it was not significant, he scoffed: "There hasn't been a spot on the reservation where they haven't been reporting damage and asking for control. They've been asking about the 1080, wondering where the poison baits are that used to be put out every year, it doesn't make sense to them."

How many sheep *were* there on the reservation, I asked Davis? "We have no count," he answered. "That's one area where the BIA has been derelict in its duties. You can come up with an approximate figure: we sheared 3.2 million pounds of wool last year; you get 7 or 8 pounds per animal." (I did the arithmetic later. This indicated a total of between 400,000 and 457,000 sheep. Coyotes reportedly had killed 2,000 sheep, or ½ of 1 percent, or less, of the total. If the figures told the truth, what was the fuss about, I wondered?)

Finally I mentioned to Davis that I understood that because of the overgrazed condition of much of the reservation, a reduction in overall sheep numbers was being discussed. If so, why bother to kill coyotes at all? Why not let them help with the reduction? "That would hardly be a scientific solution," he answered. And I expected much the same attitude from Kenneth Foster, whom I found in his office at the Navajo Parks and Recreation building. Foster began trapping in 1959 and now supervises a field force of twelve men. In 1975, using mainly traps and guns, Navajo Predator Animal Control men took 150 bobcats, 508 coyotes, 227 foxes, 1,071 dogs, 39 badgers, 48 porcupines, 2 raccoons, and 46 skunks.

"We get complaints every day about coyotes and dogs," Foster said. "All these slips on the spike," he pointed to a stack on his desk, "are current ones that we have to do something about. And with just twelve men for a reservation of 16 million acres, we get spread pretty thin. The coyotes know when lambing season starts. They follow the sheep and cut down any that are left outside the corral. And it's the same with cattle."

But then Foster's tone changed. "The people on the reservation don't want the coyote wiped out—just reduced to where he causes less damage. We figure he was created for a purpose just like all animals. People are always telling me, 'Don't kill them all. Save some!' This big reservation would be a lonesome place with no animals. We *like* to have some coyotes around, and some rabbits and rodents, too."

5 Days of Whines and Poisons

FOR CERTAIN MEMBERS of Hernán Cortés's party, the landing in 1519 on the Gulf Coast of Mexico was a homecoming of sorts: the Spaniards' horses were the remote descendants of a race that had originated in the New World but had become extinct in North America at about the time the first Paleo-Indians crossed over the land bridge from Siberia.

The centaurian sight of mounted horsemen at first dismayed the Meso-Americans. Indians had preserved no racial memory of the horse. For that matter, the art of domesticating animals was not one in which they had progressed very far. They had the dog (some breeds were raised for food; others were employed by tribes of the Great Plains to pull the travois, a rig of poles used to transport teepees and household goods); they had the turkey, a bird unknown in Europe; and Andean tribes had the llama, which they bred for wool. Parrots and monkeys were sometimes kept as pets.

The conquistadores, on the other hand, brought with them not only the horse but also many other products of European animal husbandry, notably cattle and sheep. And from subjugating the native tribes they turned to the task of subjugating the native ecosystems to accommodate their herds and flocks—just as French- and English-speaking Europeans would try to do a century later. This secondary war of conquest continues today, and as any ecologist will confirm, its eventual outcome remains very much in doubt.

In the earliest days of their presence in North America, Spanish herders made the acquaintance of an animal who would contest the white man's expropriation of what he naturally considered *his* territory with a tenacity, a flexibility, a resiliency—and, I might add, with a success—never matched by Aztec, Apache, or Assiniboin. Today, some four and a half centuries after Cortés's meeting with Montezuma and a like span of time after the first fateful, and most likely fatal, encounter of European sheep with native American wild dog, the coyote remains untamed and unconquered.

The first printed description of the coyote appeared in a Latin work by Francisco Hernández, *Nova Plantarum, Animalium et Mineralium Mexicanorum Historia*, published in Rome in 1651. "The Coyotl," Hernández wrote,

> which certain people think to be the Spanish fox, others the Adipus, and which others regard as a distinct species, is an animal unknown to the Old World, with a wolf-like head, lively large pale eyes, small sharp ears, a long, dark, and not very thick muzzle, sinewy legs with thick crooked nails, and a very thick tail. Its bite is harmful. In short, it approaches in appearance our own fox, to the genus of which it will probably be compared. It is midway between this and the wolf in size, being twice as large as our fox, and smaller than a wolf. It is said to attack and kill not only sheep and similar animals but also stags and sometimes even men. It is covered with long hair, dark and light mixed with one another. It is a keen hunter, like the fox in its ways.
>
> It is a persevering revenger of injuries and, remembering prey once snatched from it, if it recognizes the thief days afterward it will give chase. Sometimes it will even attack a pack of its own breed and if possible bite and kill them. And it may avenge an injury and exact a penalty from some troublesome man by finding out his dwelling place with great perseverance and care and killing some of his domestic animals. But it is grateful to those who do well by it and commonly signifies its good-will by sharing a bit of prey. Looking to its medical value, they say that the pain of extracted teeth may be allayed with the tail of a coyotl. The animal inhabits many regions of New Spain, particularly those tending toward cold and chill climate. Its food consists of weaker animals, maize and other kinds of corn, and sugar cane whenever it finds some. It is captured with traps and snares, and killed with the arrow.

Although a mixture of fact and fancy, this description is rather more accurate than most others drawn by Spanish pens. As Dobie put it: "For centuries after the Spanish became familiar with the coyote and with the animal's impression on native minds, their accounts of it were, aside from physical descriptions, little more than collections of native credulities." Inaccuracies, perhaps occasioned by some confusion with the swift fox, crept into the first semiscientific account of the coyote in English: Lewis and Clark's journal entry for May, 1805 in which they described "the small wolf or burrowing-dog of the prairies."

It was the biologist Thomas Say, who accompanied the Long Expedition from Pittsburgh to the Rocky Mountains during 1819-1820, who gave the coyote its scientific designation *Canis latrans*, "barking dog." He was impressed with the "wonderful intelligence of the species" and its cleverness in avoiding traps, and he wrote: "Prairie wolves are by far the most numerous of our wolves, and often unite in packs for the purpose of chasing deer."

Fact and fancy blended again in an account written two centuries after Hernández's work was published, but in this one humor rather than credulity caused the author to stretch the facts. In 1861 Mark Twain accompanied his brother, Orion, the newly appointed secretary of the territory of Nevada, on a stagecoach journey from St. Joseph, Missouri, to Carson City. A little west of Fort Kearney, Nebraska, Twain saw his first coyote. His short chronicle of that encounter, which appeared in *Roughing It*, is worth quoting in full not only because it is so entertaining but also because of the insight it gives into the then-prevalent "Anglo" attitude:

Another night of alternate tranquillity and turmoil. But morning came, by and by. It was another glad awakening to fresh breezes, vast expanses of level greensward, bright sunlight, an impressive solitude utterly without visible human beings or human habitations, and an atmosphere of such amazing magnifying properties that trees that seemed close at hand were more than three miles away. We resumed undress uniform, climbed atop of the flying coach, dangled our legs over the side, shouted occasionally at our frantic mules, merely to see them lay their ears back and scamper faster, tied our hats on to keep our hair from blowing away, and leveled an outlook over the world-wide carpet about us for things new and strange to gaze at. Even at this day [the account was written years later] it thrills me through and through to think of the life, the gladness, and the wild sense of freedom that used to make the blood dance in my veins on those fine overland mornings!

Along about an hour after breakfast we saw the first prairie-dog villages, the first antelope, and the first wolf. If I remember rightly, this latter was the regular *coyote* (pro-

nounced ky-o'-te) of the farther deserts. And if it *was*, he was
not a pretty creature or respectable either, for I got well
acquainted with his race afterward, and can speak with confidence. The coyote is a long, slim, sick and sorry-looking
skeleton, with a gray wolfskin stretched over it, a tolerably
bushy tail that forever sags down with a despairing expression
of forsakenness and misery, a furtive and evil eye, and a long,
sharp face, with slightly lifted lip and exposed teeth. He has a
general slinking expression all over. The coyote is a living,
breathing allegory of Want. He is *always* hungry. He is
always poor, out of luck, and friendless. The meanest creatures despise him, and even the fleas would desert him for a
velocipede. He is so spiritless and cowardly that even while
his exposed teeth are pretending a threat, the rest of his face
is apologizing for it. And he is *so* homely!—so scrawny, and
ribby, and coarse-haired, and pitiful. When he sees you he
lifts his lip and lets a flash of his teeth out, and then turns a
little out of the course he was pursuing, depresses his head a
bit, and strikes a long, soft-footed trot through the sagebrush,
glancing over his shoulder at you, from time to time, till he is
about out of easy pistol range, and then he stops and takes a
deliberate survey of you; he will trot fifty yards and stop again
—another fifty and stop again; and finally the gray of his
gliding body blends with the gray of the sagebrush, and he
disappears. All this is when you make no demonstration
against him; but if you do, he develops a livelier interest in his
journey, and instantly electrifies his heels and puts such a
deal of real estate between himself and your weapon that by
the time you have raised the hammer you see that you need a
Minié rifle, and by the time you have got him in line you need
a rifled cannon, and by the time you have "drawn a bead" on
him you see well enough that nothing but an unusually long-
winded streak of lightning could reach him where he is now.
But if you start a swift-footed dog after him, you will enjoy it
ever so much—especially if it is a dog that has a good opinion
of himself, and has been brought up to think he knows something about speed. The coyote will go swinging gently off on
that deceitful trot of his, and every little while he will smile a
fraudful smile over his shoulder that will fill that dog entirely
full of encouragement and worldly ambition, and make him
lay his head still lower to the ground, and stretch his neck
further to the front, and pant more fiercely, and stick his tail
out straighter behind, and move his furious legs with a yet

wilder frenzy, and leave a broader and broader, and higher and denser cloud of desert sand smoking behind, and marking his long wake across the level plain! And all this time the dog is only a short twenty feet behind the coyote, and to save the soul of him he cannot understand why it is that he cannot get perceptibly closer; and he begins to get aggravated, and it makes him madder and madder to see how gently the coyote glides along and never pants or sweats or ceases to smile; and he grows still more incensed to see how shamefully he has been taken in by an entire stranger, and what an ignoble swindle that long, calm, soft-footed trot is; and next he notices that he is getting fagged, and that the coyote actually has to slacken speed a little to keep from running away from him—and then that town dog is mad in earnest, and he begins to strain and weep and swear, and paw the sand higher than ever, and reach for the coyote with concentrated and desperate energy. This "spurt" finds him six feet behind the gliding enemy, and two miles from his friends. And then, in the instant that a wild new hope is lighting up his face, the coyote turns and smiles blandly upon him once more, and with something about it which seems to say: "Well, I shall have to tear myself away from you, bub—business is business, and it will not do for me to be fooling along this way all day"—and forthwith there is a rushing sound, and the sudden splitting of a long crack through the atmosphere, and behold that dog is solitary and alone in the midst of a vast solitude!

It makes his head swim. He stops, and looks all around; climbs the nearest sand mound, and gazes into the distance; shakes his head reflectively, and then, without a word, he turns and jogs along back to his train, and takes up a humble position under the hindmost wagon, and feels unspeakably mean, and looks ashamed, and hangs his tail at half-mast for a week. And for as much as a year after that, whenever there is a great hue and cry after a coyote, that dog will merely glance in that direction without emotion, and apparently observe to himself, "I believe I do not wish any of the pie."

The coyote lives chiefly in the most desolate and forbidden deserts, along with the lizard, the jackass rabbit, and the raven, and gets an uncertain and precarious living, and earns it. He seems to subsist almost wholly on the carcasses of oxen, mules, and horses that have dropped out of emigrant trains and died, and upon windfalls of carrion, and occasional legacies of offal bequeathed to him by white men who have

been opulent enough to have something better to butcher than condemned Army bacon. He will eat anything in the world that his first cousins, the desert-frequenting tribes of Indians, will, and they will eat anything they can bite. It is a curious fact that these latter are the only creatures known to history who will eat nitroglycerin and ask for more if they survive.

The coyote of the deserts beyond the Rocky Mountains has a peculiarly hard time of it, owing to the fact that his relations, the Indians, are just as apt to be the first to detect a seductive scent on the desert breeze, and follow the fragrance to the late ox it emanated from, as he is himself; and when this occurs he has to content himself with sitting off at a little distance watching those people strip off and dig out everything edible, and walk off with it. Then he and the waiting ravens explore the skeleton and polish the bones. It is considered that the coyote, and the obscene bird, and the Indian of the desert, testify their blood kinship with each other in that they live together in the waste places of the earth on terms of perfect confidence and friendship, while hating all other creatures and yearning to assist at their funerals. He does not mind going a hundred miles to breakfast, and a hundred and fifty to dinner, because he is sure to have three or four days between meals, and he can just as well be traveling and looking at the scenery as lying around doing nothing and adding to the burdens of his parents.

We soon learned to recognize the sharp, vicious bark of the coyote as it came across the murky plain at night to disturb our dreams among the mail sacks [in the coach]; and remembering his forlorn aspect and his hard fortune, made shift to wish him the blessed novelty of a long day's good luck and a limitless larder the morrow.

Although tempered in his case by humor and even a measure of sympathy, Mark Twain's contempt for the coyote was typical of his times. For every observer who praised the coyote's intelligence, who saw grace in his form and beauty in his song, there were a hundred who reviled him as sneaky, scrawny, thieving, and cowardly. But compare, if you will, Twain's description of the coyote with his portrait, also in *Roughing It*, of the "Goshutes" (Gosiutes), Shoshonean-speaking relations of the Utes and Paiutes who lived west of the Great Salt Lake. He knew and understood them no better than he did the coyote, and wrote about them with equal contempt—indeed the two portraits are almost interchangeable, down to the very use of adjectives. He thought the Gosiutes "the wretchedest type of mankind I have ever seen . . . small, lean, 'scrawny' creatures . . . a silent, sneaking, treacherous-looking race . . . hungry, always hungry, and yet never refusing anything that a hog would eat, though often eating what a hog would decline; hunters, but having no higher ambition than to kill and eat jackass rabbits, crickets, and grasshoppers, and embezzle carrion from the buzzards and coyotes. . . ."

Exaggeration, of course, is a tool-in-trade of the satirist, but it can be carried to extremes. Mark Twain was basically empathetic towards man and beast, but in his annoyance with those who romanticized the Indian as the "Noble Red Man," he painted a caricature that must have delighted those of his contemporaries who held that "the only good Indian (or coyote!) is a dead one." Only at the end of his tirade does he seem to suffer a pang of conscience, and he closes it, as he did the description of the coyote, on a note of sympathy: "If we cannot find it in our hearts to give those poor naked creatures our Christian sympathy and compassion, in God's name let us at least not throw mud at them." (His remorse, I might note, was not so severe that he forebore from sending the libelous paragraphs to the printer!)

Twain's irreverence towards coyotes and "Goshutes" was matched by his irreverence for the arid land they inhabited: "the most desolate and forbidden deserts," "the most rocky, wintry, repulsive wastes that our country or any other can exhibit." But this attitude prevailed among paleface pioneers from the landing at Plymouth Rock to the founding of Phoenix. People whose sense of beauty in nature and of the fitness of things had been formed amid the tidy hedgerows of Devon or the heathered moors of Scotland were awed and appalled by the North American wilderness, and could not rest until they had recast, to the extent their means permitted, the visage of their new land into a likeness of their old. Forests were for clearing, prairies were for ploughing, rivers for damming, deserts for irrigating. Lands not yet subjugated for the "beneficial use" of Western man were mere "wastes," to use Mark Twain's term, "infested" (not just inhabited) by savage tribes and savage beasts. Prior rights of man and beast were neither respected nor recognized in the drive to turn the woodlands of the Northeast into a new England, or the deserts of Utah into a new Zion. As for the wolf, or cougar or wildcat or bear, well, the wolf had been extirpated from the British Isles and so must he be from his North American domain. A 1648 Massachusetts law offered any Englishman a 30-shilling reward for the head of a wolf; Indians got 20 shillings.

There were exceptions, but all too often the westering pioneer, for all his admirable virtues, deserved the description John Joseph Mathews gave of him in his account of the Osage Indians, *Wah' Kon-Tah: The Osage and the White Man's Road* (© 1932, University of Oklahoma Press):

> . . . the ubiquitous white man, in his inscrutable desire to proclaim his presence, slaughtered wild life. The great stretches of prairie and the wild blackjack hills seemed to inspire in him consciousness of his inferiority, and he shouted his presence and his worth to the silent world that seemed to ignore him.
>
> Where the Indian passed in dignity, disturbing nothing and leaving Nature as he had found her, with nothing to record his passage except a footprint or a broken twig, the white man plundered and wasted and shouted; frightening the silences with his great, braying laughter and his cursing. He was the atom of steam that had escaped from the pressure of the European social system, and he expanded in this manner under the torch of Liberty.

An exercise in futility, still it is fascinating to speculate about what the United States would be like today if our forefathers had possessed the ecological knowledge available to us and a set of values that would have spurred them to apply that knowledge. I am sure this would be a far less ravaged land, richer in unspent resources, juster to man and beast, more varied and more beautiful, with a population perhaps half of what we number today. The fact is, however, that our forefathers did *not* possess that knowledge, and to the extent that they felt they needed ethical justification for their acts they relied on the doctrine of Manifest Destiny and a too-literal interpretation of the biblical behest, "Be fruitful, and multiply, and replenish the earth, and subdue it: and have dominion over the fish of the sea, and over the fowl of the air, and over every living thing that moveth upon the earth,"—forgetting all too often that one can subdue by means other than rape, and that dominion without stewardship is tyranny.

Take but one example: overgrazing. The deserts of the Near East and North Africa (once the granaries of the Roman Empire), the barren mountains and high plains of Greece and Spain, attest to the devastating effects of overgrazing by sheep and goats and accompanying deforestation. These object lessons from the Old World should have guided our ways in the New, but they did not. In the second half of the 19th Century ranchers began to herd their large flocks of sheep—"hoofed locusts," John Muir called them— onto the public lands of the West.

More than a quarter of a century ago, Dobie deplored the devastation of millions of acres of public lands and warned:

> Unless long-term public good wins over short-term private gain and ignorance, vast ranges, already greatly depleted, will at no distant date be as barren as the sheep-created deserts of Spain. Metaphorically, the sheep of the West eat up not only all animals that prey upon them—coyotes, wildcats and eagles especially—but badgers, skunks, foxes, ringtails and others. The surface of the earth does not offer a more sterile sight than some dry-land pastures of America with nothing but sheep trails across their grassless grounds. The free-enter- prisers of these ranges, many of them public-owned, want no government interference. They ask only that the government maintain trappers, subsidies on mutton and wool, and tariffs against competitive importations.

Is the situation better today? From what U.S. government officials have told me, I have to conclude: not all that much. Vast areas of public (and private) lands in the West are still grossly overgrazed by cattle and sheep, according to Raymond D. Mapston, wildlife specialist for the Bureau of Land Management in New Mexico: In New Mexico itself, he said, "Range conditions are much worse than they should be. We have a long way to go! Ranching operations are going to be paying for decades of over- grazing and poor management. They've brought a lot of their prob- lems on themselves. I was talking to one of our guys the other day, who comes from a ranching background. He told me, 'We never had predator trouble on the ranch unless the range was in bad condition.' It makes sense. Good healthy grassland leads to a good variety of natural prey species for the coyote and to good, healthy livestock."

This opinion was shared by Donald S. Balser, chief of Predator Damage Research at the Denver Wildlife Research Center: "We have indications that predation losses may be more a function of sheep densities than coyote densities," he told me. "When you overload the food base, you run into trouble."

William W. Rightmire, the Fish and Wildlife Service's state

supervisor for Animal Damage Control in Arizona, said to me in Phoenix: "When I see a ranch as bald as this desk, with skeletal cattle scrounging around on it—what some people have described as 'walking carrion,' I think coyotes may be doing them an act of mercy. Up on the Navajo-Hopi Joint Use Area north of here, the range is 300 percent overused!"

On a warm, sunny February day, I accompanied the Fish and Wildlife Service's Robert Roughton, who was working on coyote census techniques, on a scent post survey at a wildlife refuge south of Albuquerque. To reach the refuge we had to cross the Rio Grande on a narrow plank bridge. Halfway across we found a coyote, lying dead on the roadway against the right-hand guard rail. We got out of the truck to pick him up—a well-furred young animal in the prime of condition. Bob turned him over, and pointed to a wound on his side. "Some driver hit him on purpose," he said disgustedly. "The poor guy was caught on the bridge with no way to escape."

Once inside the refuge, we started running the survey lines. Each line consisted of fifty scent stations located immediately adjacent to the dirt road and .3 miles apart. Each station comprised a little toadstool-like perforated plastic capsule, filled with a granular fermented egg attractant, set on a small stick in the center of a three-foot circle of sifted earth or sand. At each station Bob carefully studied the bare soil around the capsule and identified if he could any animal tracks, noting the results on a pad: rabbit, blank, kit fox, question mark, blank, mouse, coyote, blank, kit fox, and so forth. I was fascinated by this, but I was conscious, at the same time, of the beauty of this land: the wide expanse of unpolluted sky, the spaciousness of the valley of the Rio Grande, the backdrop of the Los Pinos and Manzano mountains to the east, the clean smell of the westerly breeze. Bob, too, was conscious of this beauty, but at one point he looked about him with a closer focus and more experienced eye than mine. Pointing to the spare vegetation where early in this century grass grew as high as a horse's belly, he said: "A few years ago, before we acquired it, this was part of a big ranch. But the owners really skinned it! When I see land that has been abused like this, I say thank God for the federal government that now controls it!"

And yet, government ownership is *per se* no guarantee against abuse. Nor will it be until we have the courage to demand and enforce sane limitations on the densities of sheep and cattle, and the imagination to devise ways to ease the livestock industry's transition to ecologically responsible practices.

In the half century that followed Mark Twain's journey to Carson City, sheepmen, cattlemen, and farmers extended their war against wolves and other predators westward to the Rocky Mountains and beyond. Through their own efforts, those of hired "wolfers," and sometimes with the help of state money and manpower, they did a thorough job of thinning out the large populations of wolves, mountain lions, bears, bobcats, and coyotes that posed a threat to their herds. As stockmen across the West became better organized and more skilled in the use of traps, guns, and poisons, they intensified their efforts. In a September 1974 *Audubon* article, "Travels and Travails of the Song-Dog," George Laycock wrote, "In 1907 a record kill was reported for the national forests—1,800 wolves and 23,000 coyotes. Meanwhile, ranchers were exerting pressure to get the federal government into the predator-killing business. In 1914 Congress authorized an expenditure of $125,000 for the control of wolves, and the die was cast. Sheepmen then wanted the program extended to the killing of coyotes and quickly won their point. By mid-1916 much of the Western range country had been efficiently organized into predator control districts; federal trappers and supervisors were assigned to them; and a virile new government agency—the Biological Survey's Branch of Predator and Rodent Control [as it eventually came to be known]—had made the old-time free-lance bounty hunters obsolete. Within 30 years the federal program would claim 1,792,915 coyotes."

The mentality prevailing among the Survey's "gopher chokers" was clearly illustrated in letters sent to the trappers from Phoenix headquarters. Aubrey Stephen Johnson, southwestern representative for Defenders of Wildlife, showed me a sheaf of these letters spanning a period from December 1921 through January 1934. All carried the same Biological Survey letterhead, and the following letter is representative.

HUNTERS' NEWS LETTER   Phoenix, Arizona
ARIZONA DISTRICT.   December, 1923

The Hunters' News Letter is gotten out for the information of the hunters in order to convey news to them from other parts of the State, to show what each man is doing and to give general instructions from this office when necessary.

December was not a good month for predatory animal work in the State of Arizona. There are two men on the list showing a no catch record. Several others have taken but very few animals, some of them are new men and we are finding it necessary to drop them from the service and replace them with other men. If the animals are on your district to be gotten we feel that you should get them and I do not feel that it is necessary for any man to work any length of time without catching some predatory animals. . . .

For the benefit of new men on the force I wish to state that we grade their catches in the following manner: One fox is one-half point, one coyote or one bobcat is one point, one bear is ten points, one lion 15 points, one wolf 15 points, and wild dogs are graded according to the damage they have been doing or the locality they are working in. Some of them are as bad as wolves, others are not. It is necessary to have fifteen points or one-half point per day for the time you work in order to get on the honor roll. . . .

The Honor Roll is as follows:

W. A. Knibbe   Amado   31 days—4 bobcats, 3 coyotes, 2 wild dogs (poison)
J. L. Fredrick Camp Verde   31 days—1 lion, 7 coyotes, 1 bobcat (poison)
J. F. Goswick   Camp Verde   20 days—1 lion (dogs)
W. E. Morgan   Wickenburg   31 days—9 coyotes, 6 bobcats (poison)

[The Honor Roll continues.]

REMEMBER OUR SLOGAN, "BRING THEM IN REGARDLESS OF HOW."

Respectfully submitted
K. P. PICKRELL   M. E. MUSGRAVE
Assistant Inspector   Predatory Animal Inspector

A serious voice of dissent from this policy of mass extermination was not heard until 1930 when the American Society of Mammalogists raised objections at a symposium in New York. Speaking for his colleagues, Dr. Charles C. Adams, director of the New York State Museum, announced, "As field naturalists of many years experience . . . we strongly protest the excessive expenditure of public funds for drastic reduction of predatory animals in advance of satisfactory proof of the soundness of that policy. Elimination of these animals is almost certain to be followed by excessive increase of mice, rabbits and other rodents. . . . This will necessitate continual appropriations for the government poison squad. We urge provision for comprehensive scientific study *first*, as the only basis for intelligent control."

In her splendid book, *Must They Die? The Strange Case of the Prairie Dog and the Black-Footed Ferret*, Faith McNulty describes with some irony what happened next:

After the symposium events took an unexpected twist. Congress held up not only the appropriation for killing predators but *all* funds from the Biological Survey. Officials of the Survey pleaded with the American Society of Mammalogists to relax its opposition, in order not to destroy the entire Survey, with its wildlife refuges and research programs. In return, they promised to reduce their killing program and to refine their methods to minimize needless destruction of animal life. The mammalogists relented to the extent of sending Congress a message saying that their criticisms applied only to predator control, and the appropriation was restored. The chief of PARC [Predator and Rodent Control] was replaced, the work force was reduced, and a new start was made. To the indignation of the mammalogists who are veterans of this skirmish, officials of the Survey and its successor Bureau misrepresented the message that saved them as constituting approval of predator control by the American Society of Mammalogists.

Among other things this episode made clear was that PARC's activities rested on rather shaky legal grounds, since they derived entirely from a fifteen-year-old congressional directive to kill wolves. Those in favor of control hastened to shore up PARC's position. In 1931, Congress passed an act directing the Secretary of Agriculture to conduct campaigns for the destruction of mountain lions, wolves, coyotes, bobcats, prairie dogs, gophers, ground squirrels, jackrabbits, and other animals that were "injurious" to agriculture, horticulture, forestry, husbandry, game, or domestic animals, or that carried disease. Thus, any wild animal that might be considered injurious by practically anyone for any reason was on the official death list. The only limit to the execution of the act was money. Funds to carry it out would have to be appropriated anew each year.

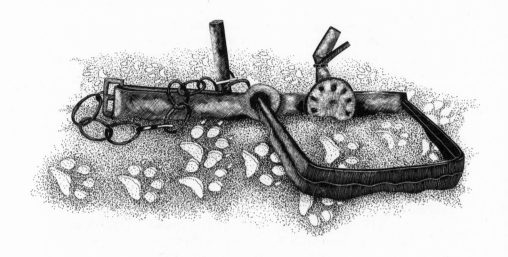

This limitation was less than paralyzing. From 1940 (when the Biological Survey was transferred from the Department of Agriculture to the Department of the Interior and was renamed the United States Fish and Wildlife Service) to 1950, the money spent on predator control went from $2,714,023 to $4,629,053. And even this near-doubling of funds, as McNulty points out, was probably not a true measure of PARC's increase in effectiveness. For in 1944 a remarkable new weapon, sodium monofluoroacetate, was added to its lethal arsenal.

Up to that point the principal "tools" (to use the jargon of the trade) available for the "taking" (another euphemism) of predatory animals had been the steel trap (inhumane, especially if not visited daily; nonselective of its victims except in the most expert hands and often counterproductive in releasing peg-legged animals that tend to become stock-killers by necessity); the rifle (as selective as the finger on the trigger, and humane if the shot is accurate); denning (digging coyote pups out of their dens and clubbing them to death, or gassing and incinerating them inside the den); the "coyote-getter" or cyanide gun (a device that ejects a load of cyanide into the mouth of whatever animal pulls the scented bait wad; nonselective but relatively humane in that death usually is almost instantaneous); strychnine (which causes a violent, convulsive death; totally nonselective—millions of drop baits, balls of strychnine-treated suet, were broadcast from airplanes over the landscape, killing any animals that chanced to pick them up); and thallium, perhaps the grimmest of all (a tasteless, odorless metallic element, so deadly that handlers must wear masks, it condemns most animals that ingest it to a slow, painful death; before they die, many go blind and lose all their hair, toenails, and teeth).

Of course, as a Fish and Wildlife official said to me, "there is no absolutely nice way to kill an animal." Perhaps some device could be developed that would inject into the target animal a lethal overdose of morphine, but to my knowledge no one has yet thought of that. It was not long after PARC began, in the mid-forties, to stud the western rangelands with 1080-poisoned baits that a violent controversy erupted over its relative humaneness or inhumaneness. Sodium monofluoroacetate, or Compound 1080, is a white, odorless, powdery salt, essentially tasteless, highly soluble in water. It has no known antidote. Like other monofluoroacetates, it is chemically stable due to the strength of the carbon-fluorine bond. It is absorbed through the gastrointestinal tract, open wounds, mucous membranes, and the pulmonary epithelium. In *A Review of Sodium Monofluoroacetate (Compound 1080)*, Stephen P. Atzert explains that "eventually the breakdown in intracellular processes caused by fluorocitrate results in the appearance of gross organ or organ system disorders. Death may result from: (a) gradual cardiac failure or ventricular fibrillation; or (b) progressive depression of the central nervous system with either cardiac or respiratory failure as the terminal event; or (c) respiratory arrest following severe convulsions."

All who have witnessed the death of an animal that has ingested 1080 have testified that it is a grisly sight indeed. After a latent period of two hours or more, the symptoms appear. Montana rancher and State Senator Arnold Rieder described, in a newspaper article, a coyote's death from 1080: ". . . a frenzy of howls and shrieks of pain, vomiting and retching as froth collects on his tightly drawn lips. . . . A scant six to eight hours after eating his meal, Mr. Coyote is breathing his last, racked by painful convulsions, [dying from] the most inhumane poison ever conceived by man. . . ."

And yet 1080's apologists, notably Dr. Walter E. Howard, vertebrate ecologist at the University of California in Davis, have maintained that it is a *humane* poison: according to Howard most, if not all, animals dying of 1080 poisoning have lost consciousness before they begin their death throes. Champions of 1080 base their defense of its humaneness on the accounts of a few humans who survived and remembered no pain, only alarm and anxiety, after being accidentally poisoned by this substance.

Of course no one can state positively that two very distantly related species such as coyote and man, with different metabolisms and even more different psychologies, will react in the same way or with equal intensity to a given stimulus, pleasant or unpleasant. Perhaps the agony of a 1080 death is more apparent than real, but many animals exhibited their gruesome death throes while 1080 was being tested at the Denver Wildlife Research Laboratory, and countless more did so out on the range, before the first human absorbed a nonlethal dose of the stuff and lived to report that the experience had not been unbearable.

So it was not humaneness that convinced PARC that Compound 1080 was the ideal tool for coyote control. Sodium monofluoroacetate had other attractions. It was cheap, and tiny amounts were effective: all you needed was sixteen grams, costing twenty-eight cents, to treat 1,000 pounds of horsemeat, or enough, theoretically, to kill 11,428 coyotes at 1.4 ounces of bait meat per lethal dose. It was easy to use: you injected it with a syringe into a 50-pound bait (a sheep carcass or a portion of a horse), dumped the bait out on the range in the fall, and left it there to do its work through the winter. It was stable: what was left of the bait when you picked it up in the spring (*if* you picked it up as you were supposed to), was just as lethal as when originally put out. Cheapness, effectiveness, stability—these were the properties of 1080 that aroused the enthusiasm of the men at PARC.

And these were the properties that so frightened conservationists—particularly the stuff's stability. This, wrote Faith McNulty, is its "most disastrous property . . ."

It does not break down in the body of its victim, and this means that any animal or bird that feeds on the carcass of a 1080 victim may be poisoned, and its body may become another lethal bait. Furthermore, the dying animals vomit deadly doses of undigested meat, attractive to many animals and birds, wherever they go. Almost all carnivorous birds and small carnivorous mammals will eat carrion, so the possibilities of this chain reaction are extensive.

Even Bureau [of Sport Fisheries and Wildlife—PARC's parent agency within the Fish and Wildlife Service] biologists acknowledge that this is a "weakness" of 1080, but they maintain that conservationists exaggerate it. Advocates of 1080 point instead to the fact that different species differ in their susceptibility to the poison. Canines are particularly susceptible. In general, birds are more resistant to 1080 than mammals. Critics of 1080 say that these variations in tolerance are a distinction without a difference, because all our native meat-eating animals and birds are capable of eating enough coyote bait to kill them. Eagles have been killed by 1080 put out for coyotes, and California condors, which conservationists are trying desperately to save, have died from eating ground squirrels poisoned by 1080.

Shortly before the advent of 1080, the sheep industry in the United States began a decline which has continued steadily (except for a leveling off in the 1950s) to this day. From 49.3 million in 1942, the number of stock sheep in the United States declined to 13.9 million in January 1974. From 34.3 million in 1940, stock sheep in the seventeen Western states plunged to 9.2 million in 1976. Profits also declined, and one response to this employed by an increasing number of sheepmen was to dispense with the hiring of herders, and to turn out their flocks in fenced pastures, unguarded by man or dog. Such sheep naturally were more vulnerable to attacks by coyotes—as they were to such sheep killers as storms, birth difficulties, toxic plants, illness, injuries, and congenital stupidity. Whatever the actual cause of loss, the coyote usually took the blame, as evidenced by an increase in predator control efforts.

Highly vocal, well-organized through their national and state woolgrowers associations, and with a political clout in state capitols and congressional delegations quite out of proportion to their numbers or financial net worth, sheepmen clamored continually for more and more poison. And PARC was happy to oblige; indeed many of its agents actively proselytized for business. Just how many coyotes and other wildlife succumbed to 1080 and other control tools never will be known—PARC was not consistent in recording the numbers of animals they knew were dead, and of course they could not record the many more who died but whose bodies were not found. But in the single year 1963, the admitted take by PARC was 842 bears, 20,780 lynx and bobcats, 89,653 coyotes, 294 mountain lions, 2,779 wolves (most of them red wolves, now on the endangered-species list), 6,941 badgers, 1,170 beavers, 24,273 foxes, 7,615 opossums, 6,685 porcupines, 10,078 raccoons, 19,052 skunks, and 601 "miscellaneous" victims. No figure was available for eagles and other birds of prey.

Not surprisingly in view of the massiveness of this effort, a growing number of people, including qualified wildlife biologists as well as lay nature-lovers, began to wonder what all this killing was about and to demand some justification. Their concern stirred action in Washington. In 1963, Representative John D. Dingell, of Michigan, introduced a bill to strip PARC of much of its money and manpower and restrict its activities to an advisory service on animal-damage problems. Secretary of the Interior Stewart L. Udall appointed the Advisory Board on Wildlife Management to study and report on PARC's activities. This group of wildlife experts, headed by Dr. A. Starker Leopold, of the University of California, turned in its report, *Predator and Rodent Control in the United States*, in March 1964.

Those who had feared—or hoped—that the Leopold committee report would be a whitewash of PARC were amazed for, although soberly worded, the report was a scathing denunciation of the whole poisoning program. "In a frontier community," it began, "animal life is cheap and held in low esteem. Thus it was that a frontiersman would shoot a bison for its tongue or an eagle for amusement . . . even today in the remaining backwoods the maxim persists that the only good varmint is a dead one.

"But times and social values change. . . . The large carnivores in particular are objects of fascination to most Americans, and for every person whose sheep may be molested by a coyote there are perhaps a thousand others who would thrill to hear a coyote chorus in the night. Control programs generally fail to cope with this sliding scale of values. Particularly when professional hunters are employed, control tends to become an end in itself, and following Parkinson's law, the machinery for its accomplishment can easily proliferate beyond real need. . . .

"It is the unanimous opinion of this Board that control as actually practiced today is considerably in excess of the amount that can be justified in terms of total public interest. As a consequence, many animals which have never offended private property owners or public resource values are being killed unnecessarily."

The Board deplored the proliferation of overlapping state, county, and local control programs, some of which included the "obsolete practice" of bounty payments, and all of which, added to the federal program, contributed to the overkill of wildlife.

On open sheep-raising areas of the western states, the Board agreed that "by far the most efficient control method for coyotes is the 1080 bait station," and that such stations, "when properly applied, according to regulations [not more than one to a township] . . . do an effective and humane job of controlling coyotes and have very little damaging effect on other wildlife.

"On the other hand," the Leopold Report continued, "we are

aware of a good many instances where regulations are not followed and where 1080 stations are placed much closer together than they should be, excessive doses of poison are used, and the poisoned bait is not always picked up in the springtime. Abuses of the regulations are condoned in some PARC districts. Under these circumstances, considerable damage can occur to other forms of wildlife as well as to domestic dogs."

The Board found "little justification for general coyote control" on rangelands occupied only by cattle: ". . . in great areas of the West, cattle and coyotes seem to live amicably together, with no reported losses whatsoever." Similarly, it found little or no scientific justification for controlling predators to protect game or other wildlife populations, and little or no scientific proof that, in outbreaks of rabies among wolves, coyotes, foxes, raccoons, or skunks, calling in PARC can "hasten the termination of the epizootic, or . . . reduce the danger to humans and domestic livestock." The distribution and use of 1080 as a poison for field rodents should be banned, the Board urged, because of the widespread secondary poisoning of unintended victims, principally carnivorous birds and mammals.

Finally, the Board made six specific recommendations:

1.   That the Secretary of the Interior appoint a permanent Advisory Board on Predator and Rodent control, [to] include carefully selected individuals representing the livestock and agricultural interests, conservation organizations, and technical organizations such as the National Academy of Sciences, American Society of Mammalogists, American Society of Range Management, and the Wildlife Society.

2.   That the Fish and Wildlife Service and its Branch of Predator and Rodent Control completely reassess its function and purpose in the light of changing public attitudes toward wildlife . . . [and that] the goal of PARC should be to control animal damage on an absolute minimum basis consistent with proven needs to protect other resources and human health. . . . [The Board warned that] unless the government control program undergoes a drastic and critical internal revision of operational objectives and procedures, an even more drastic revision will sooner or later be forced by the public, with possible serious curtailment of the control functions which we concur are locally important.

3.   That the justification for each local control program

should be documented far better than at present. . . . as a form of justification, narrative descriptions of damage should be supplemented with quantitative statistics on the true extent of damage.

4. That [the Fish and Wildlife Service undertake] a greatly amplified research program [with particular emphasis on], (1) finding more specific controls for pest species, thereby minimizing unnecessary killing of innocent animals, and (2) development where possible of repellents, fences, and scare-devices which would preclude the necessity of killing any animals. The research could well be financed out of savings resulting from curtailment of the present PARC program.

5. That [a new name be selected for PARC] that clearly connotes a broad management function.

6. That [legal controls be greatly tightened over the] distribution and use of 1080 or any other poison capable of having severe secondary effects on non-target wildlife species. . . . [and that] the legal regulation be extended to exclude the export of 1080 to Mexico, or any other foreign country, where the danger of misuse is substantial.

More than a year passed between delivery of the Leopold Report to Secretary Udall and his announcement in June 1965 of its acceptance as a "general guidepost for Department policy." In the meantime, anguished delegations of sheepmen had descended on their congressmen in Washington, pleading with them not to allow any cutback in predator control.

The sheepmen need not have been so worried, for only one, and the least important, of the recommendations was implemented: PARC was re-christened Animal Damage Control and was made part of a newly created Division of Wildlife Services whose functions included pesticide appraisal and monitoring, and "wildlife enhancement." Field men—formerly mammal control agents but now dubbed district field assistants—were given pep talks about the need to readjust to a new philosophy and were specifically enjoined from any more proselytizing. And the killing went on. During the first year following the Leopold Report, the federal predator-control program accounted for 90,236 coyotes, 17,294 lynx and bobcats plus sundry other animals, at a cost of $5.6 million. By some adroit twist of logic, Wildlife Services public relations people soon were announcing that the Leopold Report essentially had been a vindication of the federal predator-control program.

Convinced that they had been gulled, conservationists returned to the attack, and once again Representative Dingell introduced a bill to abolish federal predator control and replace it with an extension trapper advisory service modeled on the Kansas system. Dingell held hearings early in 1966, and the first man in the hot seat was Dr. Stanley A. Cain, a coauthor of the Leopold Report (formerly a Professor of Conservation at the University of Michigan School of National Resources and now, in 1966, Assistant Secretary of the Interior for Fish and Wildlife). There had not been enough time or enough money to implement some of the report's recommendations, he testified, and some others had been rejected as impractical. Nevertheless, he assured Congressman Dingell, Wildlife Services now was "very sensitive to the principal points that we want effective control with the least possible damage to the environment and other organisms and we want it only where it is demonstrably needed."

Congressman Dingell was unimpressed: "But actually you . . . have nothing to show this committee in terms of concrete changes, with the exception of revision of the internal structure of your agency? . . . Am I correct?"

Dr. Cain: "You are not completely correct, I don't believe."

Congressman Dingell: "But I am sort of correct?"

Dr. Cain: "You are essentially correct."

Among the many individuals who had written to the Department of the Interior in protest against the predator control program, was Dr. Raymond F. Bock of the Pima, Arizona, Medical Society, whose letter read, in part: "One wonders whether someone in your department has gone mad from a personal hatred of predators. . . . We have found consistent objection to your methods by trained biologists. . . . We wonder what kind of misfits may be perpetuating this campaign." Paul Maxwell, a fur trapper from Grand Junction, Colorado, supplied one answer at the hearings.

"I know personally," he told the subcommittee, "the type of person employed by the federal government in our area—I am only speaking for the western part of Colorado—to do this poisoning. They call them mammal control agents. One operates a bar in Craig, another is a building contractor. All are guides during the hunting season, and none have a guide's license. One is a drunk who will stoop to anything to get a few scalps to turn in, to show he has been doing [what he was] supposed to be doing. This includes running another trapper's line, which, incidentally, in this case was mine. . . . I could go on for hours . . . on how the Fish and Wildlife Service handles their 1080 and how they just let the sheepmen come in and help themselves to the treated baits and how they pass out strychnine pills by the box to sheepmen, and how the druggists in that country are all handling thallium poison and the sheepmen are buying it, and that's the worst poison known to man. . . . And I could tell you how the sheepmen shoot game animals, deer and antelope by the hundreds and treat them with poison on their own. I know where there is fourteen deer baits right now. . . . It would make your hair stand on end, all the things I really have seen. . . ."

Dick Randall, the retired Fish and Wildlife Service acting supervisor mentioned in Chapter 2, bears no resemblance to the men Paul Maxwell was describing. Tall, trim, with crew-cut gray hair, Dick is a soft-spoken man with a pleasant, courteous manner. Yet between 1958 and 1973 he personally executed many hundreds of coyotes and countless other animals. Now a field representative for Defenders of Wildlife, which has spearheaded a crusade against abuses in predator control, Dick makes no effort to gloss over his past.

"When I first began working for Fish and Wildlife," he said, "I was an ideal recruit. I thrived on long hours and hard work. I was very gullible. I was in awe of Ph. D.'s whose studies convinced me that my 1080 baits were a highly selective, very responsible kind of predator control. And I was an excellent trapper. The criterion for judging the worth of a trapper is the number of predatory animals he destroys. Fish and Wildlife records will confirm that I was rated among the top trappers in Wyoming, and Wyoming is noted for outstanding predator control people."

What kind of people were the trappers he worked with? I asked him. What was their attitude towards the animals they killed?

"Most of the trappers I know," he answered, "are people that have lived in the hills, have trapped all their life anyway, and like that kind of thing. And when they're paid money to do it they're in seventh heaven because they're doing something they love to do anyway. But I don't think you can pick out a trapper and say, 'this is a prototype.' No way. They're individuals. And some are responsible kinds of people and others, I kind of hate to see them running around with poison and traps. You run into some that can burn up the coyote pups in a den, and listen to them squeal, and never think a thing about it. You run into others who'll get tears in their eyes, you know, and cuss the sheepmen, and I was one of them. I just couldn't stand to see anything suffer. You run into some trappers that if their den dog doesn't behave and do just what they think he should, they'll shoot him. Others, my God, you hurt their dog or kill their dog, they'll kill *you*!"

My day-long conversation with Dick Randall offered the best "inside" look I got regarding the aspects of predator control that prompted the Leopold Report and that led, as we shall see, to a major overhaul of the system in 1972. Of course Dick was describing the situation in Wyoming, which was perhaps more extreme than in states such as New Mexico or Arizona because of the greater power and intransigence of sheep men in the more northern state. ("Wyoming sheepmen," a high Fish and Wildlife Service official said to me, "have done and are doing more harm to their industry than anything else—including the coyote.") But PARC was PARC (or, to use the new initials, ADC), whether in Montana or in Texas, and differences between states were more of degree than of form.

The demands of the job were such, Dick explained, that often a trapper could not "check his traps every day or every two days or even every week—because you may have to set them at one end of the county and you may be off in the other end working on some trouble calls and you may not get back for two weeks. So there's no way you can be responsible. The last couple of years I worked for Fish and Wildlife, I just about quit setting traps because I couldn't get back to them. I'd be up in Lincoln County, up in Sublette County, or someplace else, and then come back and look at the animals in the traps, and say, to heck with it! It got to where I wouldn't even set traps."

How did those trapped animals die? I asked Dick. Of thirst, or what?

"Oh, it depends on the time of the year. In the summer, they last two or three days in a trap, the sun will get them. In the winter, they can last a week, two weeks. If there's snow on the ground they'll lap up the snow for water, and just stay in the trap and starve to death. When you find them, why, they're just skin and bones, the ribs show through. Of course some animals will twist out of the trap, or chew their feet off, and escape.

"They say this offset trap—it's got a three-eighths-inch offset in the jaws—they say it's humane. That's nonsense. I've got lots of pictures of animals with their foot in there, their paw's all swollen. The trap shuts off circulation, it cuts to the bone, same as any other trap. And when you skin the animal you find, if it's been in there a day, there are red marks up to the shoulder and swelling and a small amount of pussing has started. In a couple or three days you get big masses of jellylike stuff and pus, and you can see the suffering the animal has gone through because that stuff goes clear up into the shoulder and over the withers. It's tough!"

But the biggest impediment to trappers' doing a professional job, Dick emphasized again and again, was the influence of the sheepmen themselves. Cooperative agreements exist in a number of states, under which predator control is financed jointly by federal funds, state funds, and livestock-industry funds raised through a mil-levy on sheep and cattle. "In Wyoming we have these Predatory Animal Boards—PAB boards—in each county, made up of five sheepmen," Dick explained. "They have the say-so over the mil-levy money, which goes to pay trappers' wages. This gives the sheepman a thumb on the trapper. They say, 'We're paying your wages,' and they *are* paying a part of the wages, 'therefore we want you to do this and that.' While I was acting supervisor, I had to lay off three trappers who were *very* good people. These people were staying around trouble call areas working on coyotes that were causing problems, and they were doing a great job—they cut predation a whole lot. But they weren't killing many coyotes per month, and the way sheepmen figure out how good a trapper is is by dividing the number of coyotes into his salary. And the cheaper the coyote, the better the trapper is. So these fellows were running up $60 to $100 a coyote a month. I had to lay these three people off, and they were replaced by a $10 bounty. Now that's the kind of thinking that governs predator control."

On another occasion while he was acting supervisor, Dick appealed to the Lincoln County PAB for more money for the three trappers he had assigned there, because "they were underpaid and, darn it, living expenses were up. The board said, 'We don't have the money, but we'll let them keep their furs. Tell them they can sell their furs and keep the money.' [Ordinarily money realized from the sale of furs from animals the trappers had caught would revert to the PAB.] The first year—it was late in the year when we got this agreement going—I think one trapper caught about $1,800 worth of fur that supplemented his salary. But the next year they were all set for this. They quit trapping coyotes, because the coyote was worth only about $20–$25 for a pelt. They trapped bobcats. One trapper trapped $5,200 worth of bobcats. Well, he wasn't doing anything about predator control, he was trapping bobcats! Coyotes—he could care less, because they weren't worth much money and the cat's a whole lot easier to catch. But, that's your thinking behind predator control. There's nothing scientific, or biologically sound behind it."

Trappers' wives, apparently, felt the pressure, too. "You go talk to my wife," Dick advised. "She is so full of sheepmen getting her out of bed at midnight and screaming and hollering that 'Coyotes are putting me out of business,' and I'm out camped somewhere and I come in just long enough to get the messages that were left and I go back out and she's plumb *full* of those people. And so are a lot of other trappers' wives." Dick's wife corroborated this statement at dinner. "Sheepmen!" she spat. "I'm sick to death of them. They're a bunch of whining crybabies. Calling me up at all hours of the night, whining and crying for more traps, more poison, more attention. 'Where's your husband? Why isn't he out here? Send him out here right away!'"

Dick added to this. "The trapper just can't go out and satisfy all these people. So sometimes he takes shortcuts. He does things that really shouldn't be done. Some trappers bend under pressure and some don't. I've got records in my files of 1080 baits, twice as many in an area as there should have been, three, four times as many. And that was partially due to pressure from the sheepmen. Because, doggone it, you're working for the sheepmen, I don't care *what* they say. Possibly it's changing some now but the sheepmen controlled the field people that put out the toxicants. And if they wanted a bait here, and another bait there, and they wanted a little 1080 for their own personal use they usually got it.

"A sheep rancher knocked on my door one morning—he had killed 14 antelope and he wanted me to treat them for him with 1080. I damned near did, too, he just about had me buffaloed. He was a big-wheel sheepman. Took me out and showed me the antelope and I finally convinced him, the carcasses were cold, he'd waited too long, it wouldn't do any good to inject them. That's the way I got out of it. He was all hot to trot."

There *were* some good, responsible trappers, Dick emphasized more than once. There also were "some *good* biologists in Fish and Wildlife, and I'm sure if the political pressure from the livestock industry were removed from these people we could have had a legitimate, sound kind of predator control. But the thinking was, the more coyotes you kill, the more bobcats you kill, the more mountain lions you kill, that's 'controlling' a species, that's 'management.' Bull manure! It just doesn't work, it never has worked. Sheep numbers have decreased all throughout the thallium years, throughout the coyote getter years, throughout the 1080 years. The only time sheep numbers in the West really bounced back was during the Korean War when wool went to as high as a dollar fifty-one a pound. Right away sheep numbers started to increase. Economics govern the number of sheep out here, not predators."

Economically, Dick went on, the industry was a nightmare. "It's *so* archaic that you wonder if it ever will adapt to the twentieth century. This is one industry that without all the subsidies it's getting—the incentive payments, the tariffs on foreign wool and mutton, the subsidized killing of competing wildlife, the experimental breeding stations that are located all over the United States and that cost the taxpayer several million a year (most industries research their own products, but the sheep industry doesn't)—if all these subsidies were ended and the industry had to operate as a real free enterprise, it would collapse overnight."

One of the problems that was not entirely of the industry's making, Dick admitted, was the scarcity of qualified labor. I asked him if sheep ranchers were still importing Basque herders. "Oh, yes," he replied, "I think we probably have half and half Basques and Mexicans—a lot of the Mexicans are wetbacks. And the Basques, especially, some of them around here are good herders. But usually they will herd for a year or two, then they'll go find something else. They'll work in the trona plants and earn a lot more money, get two days off a week, insurance, retirement, *lots* of stuff. Because who the heck wants to go out there and work seven days a week for $300 a month? You can make more than that on welfare.

"Not *too* many years ago, twenty-five or so, sheepherders didn't ride horses. They *walked*, with the old crook, you know, like the old shepherd. They would leave the wagon in the morning with the sheep, they packed their dinner and a 30–30 rifle, and they walked ahead of the sheep. In the evening they came back in with the sheep. They didn't drive them. Turned the sheep around and let them graze their way back to the wagon, none of this sicking the dogs on them to hurry them up because supper is about ready. And if they had coyote trouble they'd take a teepee and plant it out where the sheep were bedded, and they'd sleep in the teepee with the dogs. And they *knew* if the coyotes came around. The dogs would let them know, the sheep would stir, they were light sleepers. But no more. Now they gallop out in the morning and try to figure out which way the sheep went."

Then Dick gave a wry laugh. "You know," he said, "it's the only industry I know of where you've got, say, $50,000 to $100,000 worth of livestock in a herd, and you're downtown trying to drag some wino out of the bar that you've never seen before and that's supposed to be a sheepherder. And you're going to take him out

there and turn him loose with your investment, and you don't
know who he is, he doesn't fill out any papers or take any tests, all
he gives you is his social security number. And he can make or
break you!"

Regarding actual predation losses, Dick commented, "what I
found about predation is that you can't say that it's 2 percent or 5
percent or 10 percent over a whole area. It's a very local thing.
Predation occurs in one place and over the hill it *doesn't* occur, and
some sheepmen have a lot of predation, others have very little.
Part of it is due to the area they operate in, part of it is due to management of the livestock. We have a couple of sheepmen here who
have very *little* predation. They have Basque herders, they move
their camp every day, usually, at least every two days, they get the
best feed, they have the heaviest lambs, they have less predation
because they have very good herders. And they spend most of
their time out there with those sheep. Every day they're out there.
Some of the other owners get out once a week, once every two
weeks, or they have a foreman that goes out sometimes. It's a business you have to stay on top of to make any money."

"How come some of these guys will do it, some others won't?" I
asked.

"Because it's just like any other businessmen, they put in more
time taking care of their business, they're on top of it all the time,
and they have responsible people working for them, maybe they
pay them a little more money, and maybe they treat them
better. . . ."

"But the others will claim they can't afford it?"

"Right."

"*Could* they afford it?", I asked.

"Well, they'd make more money with better management. One
sheep outfit here has one herd of sheep. They probably make more
with that one herd of sheep than some of the big outfits do with
five, six, or seven herds. It's due to management. I know one of
these outfits that I think is pretty responsible, we met them
coming out of the national forest one year, and asked them how
they'd summered. They'd lost one sheep to a bear, two sheep to
coyotes, three sheep they couldn't account for. That was their
entire summer loss, and they knew what had happened to every
one of those sheep. Others would come out of the forest and leave
300 or 400 sheep up there they couldn't find, and sometimes we'd
find them when we went up in there putting out 1080 in the fall.
We'd come out and tell them, 'your sheep are here, your sheep are
there.' I doubt that anyone ever went back for them. Most of them
would stay up there and die. They're an ignorant animal, in fact I

don't think they have any brains at all. They'd get snowed in,
bunch up under a pine tree, eat off the pine bark and needles as
high as they could reach, and starve to death. It was fairly common, when we'd go back to pick up or bury what was left of the
baits in the spring, to find the bones, the remains, little groups of
them under the pine trees. And of course they're chalked up to the
coyotes!"

"Overgrazing often can be the cause of lamb deaths," Dick said.
"I know a sheep rancher out here—you wouldn't believe that guy!
He's boasted about the coyotes he's trapped, how he takes a burlap
bag, cuts out a hole for the head and two holes for the front legs,
pulls the bags on the animals, pours kerosene on, sets them on fire
and turns them loose. And he *laughs* when he tells that, as if it
were the greatest thing!

"Well this guy had a lot of lambs die out there one spring. He
was in an overgrazed area, he abuses the hell out of it, it's the biggest dust bowl in the country. That spring even fishermen were
calling me up and saying, 'What in the hell's happened to all the
lambs up there that have died?' 'It's the ewes running off and leaving the lambs,' I told them. Lots of them were yearling ewes.
Grass was twenty feet between clumps. And the ewe's interested
in finding something to eat, so she'd drop the lamb and go on about
her business. That's all it was. Well, Bill was screaming bloody
murder out there. The coyotes were putting him out of business.
There was no *way* you could convince him. Coyotes were doing it
all."

Apparently this rancher was not alone in being hard to convince.
"It's *so* difficult to communicate with any of these sheep ranchers,"
Dick told me. "When they have their mind made up it's made up.
There's no way you're going to change it. You can point out, 'This
is a bobcat track, this is not a coyote track.' He *knows* it's a coyote
track, so it's going to *remain* a coyote track!" Some men, of course,
were more reliable than others. "I know a couple of sheepmen
right now," Dick said, "who, if they told me that a skunk had got in
with the herd the night before and killed forty of their ewes I'd
believe it. They're that kind of people. Others, I wouldn't believe
them if they said they had a coyote on the ranch! You learn what
sheepherders to believe, you learn which ranchers are reliable,
but still you should check on them now and then because they can
get carried away. All claims of predation should be verified, but
often the trapper just doesn't have the time.

"Because most of the time your trouble calls are just phone calls,
some sheepman will call and whine, 'They're putting me out of
business! I just can't stand the loss. They're eating me up!' And it's

a lot of hot air. *So* many times I've had ranchers call, 'Found a coyote den, take you right out and show you!' And you drive out a hundred miles to have them point out a badger hole! A bunch of newborn badgers have tracked up the area, and they can't tell the difference between the tracks. And if they're that unobservant I can understand why they think eagles kill their lambs, why they shoot them so much. Most of these trouble calls, I'd say over 50 percent, are because a coyote howled, because a coyote was seen, because a coyote track was found in the area. And that immediately is . . ."

"Cause and effect?" I broke in.

"Yeah! But with that airplane, you can *find* dead sheep, if there's any out there. You fly in there at dawn, the sun has just cracked, and it's making that beautiful skim light which gives texture to everything, and a dead lamb lying there shines up like a million dollars. A lot of times we went into areas and found a few dead lambs when the rancher didn't have any idea he was losing lambs to coyotes. Other times you'd get trouble calls, 'they've killed twenty last night, they've killed fifty,' you'd go in and search the area with the airplane and go in on horseback and you wouldn't find any dead sheep. And if they were there you'd *see* them, there's no way you can miss them when you're dragging the sage-brush with that Supercub or a helicopter.

"But one thing we learned with that helicopter," Dick continued, "was that we could stop predation without ever killing a coyote. The rancher would phone and say, 'I lost six sheep here and I lost five lambs there,' and you'd say, 'Let's go look at them.' 'Why, I don't have the time. Too far away!', he'd say. Ahah! So you bring the helicopter in. 'Come on, it'll take two minutes, let's go find them,' you tell him. And he can't find them. And they learned that, real quick. So right away their predation dropped 50 percent or better, because they knew that if they hollered 'dead sheep' we'd want to go look at it."

Sometimes just a good bull session with the rancher would cut the losses. "We've sat down many times with some of the people that we *knew* were way over in describing how many lambs they'd lost to coyotes," Dick said. "We'd stay at the ranch, and sit there in the evening and have coffee and talk about these claims. 'Aww, you didn't lose *really* this many, did you?' 'Well, no.' And we could usually whittle them down, sometimes 50 percent, 75 percent, just by visiting with them a little bit."

"But why do they *make* these claims, if they don't believe them?", I asked.

"To get the trapper out there, to get some attention. If they don't see the trapper now and then they feel neglected. And maybe they've *got* some coyotes. Maybe they have *no* trouble. But the idea is to scream louder than the other guy, and the squeaking wheel gets the grease, you know. And of course some of your sheep ranchers are more vocal than others, more influential. Certain ones, when they holler you'd better go because they control the PAB boards and you want to take care of those people. It has nothing to do with responsible wildlife management."

Often it had more to do with psychology, Dick suggested. "Actually, much of our work wasn't helping the sheep industry any. It wasn't getting to the roots of the problem. But never mind. As long as we're killing some of the coyotes out there, it gives the sheep ranchers a psychological boost. It's one of the traditional things. In the sheep industry, regardless of whether the baits are being eaten by the coyotes, if the baits are out there on the hill there's a feeling that all is well."

"Something's being done for them?" I interjected.

"Yeah. There's poison on the range. It's killing the coyotes and things are better. Actually, the 1080, you know there's probably been a million dollars or more spent in studying 1080, in the Denver Research Center. I think there's been two small field studies conducted, aimed at determining what species of non-target wildlife are poisoned by 1080. All the research stuff comes from the lab, it's under controlled conditions. You're using 1080 that's mixed to the proper dosage, you can use ground meat, put in the right amount of 1080, tumble it and come up with the right combination. And it's true, a coyote is less resistant to it than, say, a golden eagle.

"But in the field there is no possible way to treat bait material to conform to laboratory specifications. There's a lot of reasons. The 1080 doesn't readily penetrate membranes, it has an affinity for protein which makes it clump up. Then there's no way that two operators will treat the same carcass with the same amount of poison, because one is going to push the plunger a little heavier than the other one, one is going to make more injections and disperse the poison more than the other. What you're trying to do is get sixteen one-thousandths of a gram of 1080 spread evenly through each pound of meat to make it conform with laboratory specifications. In the first place, whoever's treating the carcass doesn't have any idea how much meat there is on it.

"Now, this lamb would probably weigh fifty pounds on the hoof," Dick pointed out, showing me a slide. "You got him gutted, he's down to thirty-five, thirty-two pounds, you've got a lot of

untreatable meat, you've probably got twenty pounds of meat on
that carcass that can be treated. But you've got enough 1080 for
fifty pounds in that bottle. Well, you don't dump the solution that's
left, you put it in the carcass. So, they're overtreated two or three
hundred per cent plus you have the concentrated hot spots, which
are even worse. And you'd wind up with baits in the field that
would kill any of our carnivorous mammals and many of our
raptors, including eagles."

Around 1969, Dick said, Fish and Wildlife started putting a
tracer material in their 1080. They already had a tracer in their
strychnine cubes. These tracers could be detected by examining
the digestive tracts of poisoned animals under a fluorescent light.
Strychnine showed up green, 1080 fluoresced yellow. This was
done, he explained, "because of tort claims against the government
by people who'd lost dogs to poison. There was so much poison on
the range, not only government poison but stuff put out by individual sheepmen that had their own poison campaigns going, the government had to have some way to say, 'look this is not our poison,
somebody else put it out, you can't sue the government.'

"Well, I saw this tracer as a means of trying to determine for
myself what in heck my poison stations were killing. So I put on my
own pick-up-the-dead animal campaign. I didn't have a whole lot
of time to search for dead wildlife. Carcasses disappear fast, they
deteriorate, and they're hard to find to start with because 1080 is
fairly slow acting and the animal's not going to drop right close like
it would with strychnine, so they could be a long way away and
hard to find in the sagebrush. But I looked for dead wildlife on
horseback, on a trail bike, once in a while I would spot carcasses
from the airplane, sometimes some of my friends would tell me
they'd seen this dead here, and this dead there and I'd go out and
recover it. If I'd had five or six graduate students out there picking
up dead wildlife in the areas that I could point out to them where
the 1080 stations and strychnine were, I'm sure we would have
come up with *hundreds* of animals.

"As it was I brought in 140, 150 bodies of wildlife, some of them
whole, some just the entrails. I had two freezers full of them. And
when I'd get time I'd thaw them out, and probe through the
entrails under an ultra-violet light. Sometimes it took hours before
I could find the tracer, sometimes it was easy, like this golden
eagle," he said, projecting another slide.

I jotted down a list of poisoned animals Dick had brought in:
dog, coyote, badger, bobcat, black bear, pine marten, mink,
skunk, weasel, golden eagle, great horned owl, redtailed hawk,

prairie falcon, magpie. On one list of thirty-six autopsied birds, half showed 1080 tracer, six had strychnine tracer, twelve showed no tracer.

"But they agree," I said, "proponents of 1080 agree that yes, even if it's spaced correctly and so forth, it will kill nontarget animals, but just locally. They say that most of these nontarget species do not travel as far as the coyote, so although you may kill the fox or badger or skunk that lives adjacent to the bait, you will not get the one that lives a mile or two or five away."

"Right," Dick answered. "Of course you can't say that about raptors, because they travel all over. And most of your mountain bait stations are placed on high ridges where they're susceptible to raptors. Also, there's the secondary poisoning effect. You start out with one bait station, a coyote eats on it and goes off to die a mile or two away, and it throws up along the way, its body and its vomit become bait stations themselves. So right away you've extended and multiplied the lethal range of the original station.

"Another thing they'll tell you about 1080," he went on, "is that by killing coyotes it benefits competing predators, even though some individuals may be killed. But I would differ with that very much. We have never had any non-game mammal censuses. Only huntable species, huntable birds, huntable mammals are counted. We have no idea how many marten communities we have in Wyoming, we don't know if we have any fisher, we don't even know if we have any ptarmigan. And so when Fish and Wildlife says we're benefiting non-game species, they don't know what they're talking about. Because they don't know to start with how many of these communities there were, what the populations were, they don't know if they've increased or decreased, they haven't any idea what 1080 has done to them, whether it's helped them or hurt them, because there's been no study on which to base such comparisons.

"There used to be an old sheepman named Bill Spicer. He's dead now. He was kind of a legend in the sheep business. He used to say, when 1080 first came out it paralyzed the coyote population, and it paralyzed the trappers, and the only one that ever got over it was the coyote. Well, my records show that in 1960, in the Baggs, Wyoming area, coyotes ate about 60 percent of the 1080 baits I put out. The last 1080 records I have show that coyotes consumed about 8 percent of the 1080 stations I placed in 1971. The decrease in bait consumption didn't come from a great decrease in our coyote population. Coyotes simply learned to avoid lethal stations. I am sure that scavenger coyotes, those that fed on carrion, were the first to go. Those who killed their prey survived.

"But that's typical of the coyote. He adapts to everything. Take traps. You can usually catch the dumb pups. But if they make it to the age of two they're pretty sharp. I don't know what instinct they have but they can come in and sense you have traps there, and dig the trap out, and push it out of the hole with their nose, and turn it over, without tripping it, and leave it there for you to look at and feel like what a damn fool you are for even *trying* to catch the guy!

"Back in 1939, Fred Marlman proved that an unlikely looking contraption called a cyanide gun could slaughter coyotes by the thousands. They were inexpensive, simple to operate, so they were sown like wheat throughout coyote ranges. In a few years, a large percentage of the coyote population was either avoiding the cyanide gun or was using it for the same purpose a city dog uses a fire plug.

"If you've been around coyotes, for say the last twenty years, you can see how they've adapted to even changing their dens. They're very much harder to find, now, than they used to be. The coyotes take more pains in hiding them, getting them into areas where it's harder to track them. Lots of dens, now, are in areas like adobe hardpan where you don't see any tracks at all.

"And aerial control—they adapted so quickly to that! Aerial hunting could be one of the best means of stopping coyote trouble. Only it's not used that way. You don't only work the trouble areas, you work the areas up to 50 or 100 miles away. You keep flying and shooting coyotes as long as you have money to pay the pilot.

"But it used to be a lot easier than it is now. It's not coyote nature to run for a hole when he senses danger. He is not that kind of an animal. When we began aerial control, if we started a coyote running with the airplane he might go into a draw but he would come out the other side. And you'd get him. Especially in winter, you flew 500 feet in the air and anything out there a mile or two away that was a black speck moving was a coyote. And they didn't have a *chance*. And I killed so many coyotes I got ashamed of myself. I think I got 700 and some coyotes in three months. Of course next spring, I didn't notice *any* difference in the amount of telephone calls I got. It was still the same old whine, 'the coyotes are putting us out of business, the coyotes are eating us up.'

"But nowadays, when you are hunting in winter, you can see from the tracks that the coyotes heard the airplane. You can see where a coyote has been walking or loping, and all of a sudden he's broken into a dead run and he's headed for a draw, he's dived for a hole because he knows about that airplane. Maybe he's been burned once by some buckshot, maybe he's witnessed another coyote get killed. I don't know how they communicate, but they do. They've got this whole thing figured out. They probably can read the numbers on your plane!"

"Would it be possible," I asked Dick towards the close of our discussion, "to eradicate the coyote if we really wanted to?"

"Oh, I think it might be possible, if we had several billion dollars, and all the national parks and monuments were open to coyote killing," Dick replied. "You'd take half of our other wildlife species along with it because it would take an awful lot of poisons and traps. Even then I'm not sure. Because the coyote's got one final trick up his sleeve. That's his prolificness.

"Take an area like Yellowstone. There's no control in there, there hasn't been for quite a few years, and there aren't many coyotes. There are natural factors that keep the population in check. In fact they could have used a lot more coyotes up there last spring with the elk die-off they had due to the tough spring. There weren't enough coyotes to go around.

"But if you go into an area and start intensive control, immediately the litter size increases. And whereas in Yellowstone maybe 15 to 20 percent of the yearlings might breed, in an intensive control area you might wind up with 80 percent of the yearlings breeding. So, they've immediately answered the attack by increasing their population. You've got a more reproductive bunch of coyotes. And a more *troublesome* bunch. It's happened so many times, that we've gone into an area where there were plenty of coyotes but there was absolutely no predation on the sheep, and we've killed a few coyotes, and a few days after that all kinds of trouble broke out. Coyotes were killing sheep. We should have left well enough alone.

"The coyote has adapted in every way *possible* to survive. And he's so far ahead of you. Actually, it seems ridiculous when you stop and think about it. There's millions and millions of dollars, and there's the latest in four-wheel drives with winches, scope-sighted rifles, the very best, the latest thing in toxicants, helicopters, fixed-wing aircraft, *everything* after one poor, dumb, four-footed animal. And he's made an ass of everybody that's been chasing him. He's adapted to everything they've thrown at him, and he still survives pretty well, in some areas. It *is* kind of ridiculous, when we're supposedly *so* superior, and yet we've been unable to cope with this animal. Of course, a lot of the 'coping' never needed to be done in the first place."

This, of course, had been the conclusion of the Leopold Report. And in the climate of burgeoning environmental concern that characterized the sixties and early seventies, it also became the conviction of an increasingly large and increasingly vocal segment of the

American public, who theretofore had not cared or had not known about the mass killing of predators. This new interest was reflected in a spate of articles in popular magazines and of books addressed to the general reader, that were sympathetic to carnivores and critical of their persecutors. Among the books were Farley Mowat's *Never Cry Wolf*; Roger Caras's *The Custer Wolf*; Sally Carrighar's *Wild Heritage*; Robert Murphy's *The Mountain Lion*; L. David Mech's *The Wolf*; Russell J. Rutter and Douglas H. Pimlott's *The World of the Wolf*; Ronald Rood's *Animals Nobody Loves*; Ian McMillan's *Man and the California Condor*; Faith McNulty's *Must They Die?* (which first appeared in *The New Yorker*); and Jack Olsen's *Slaughter the Animals, Poison the Earth*. In this climate it is not surprising that the revelation that certain sheep ranchers in Wyoming and Colorado had hired a helicopter pilot to fly missions on which more than 800 eagles had been shot down created a massive furor, which culminated in Senate hearings and in the indictment of the ranchers (who wound up paying token fines).

In April 1971, the Department of the Interior and the Council on Environmental Quality, having decided it was time for another impartial review of the control program, agreed to sponsor such a study jointly. A contract was signed with the University of Michigan's Institute for Environmental Quality, and the Advisory Committee on Predator Control was appointed, made up of seven distinguished biologists including the chairman, Dr. Stanley Cain, and Dr. Leopold. The Cain Committee's report was published in January 1972, under the title *Predator Control—1971*.

In comparison with its predecessor, the Leopold Report, the Cain Committee Report was at least as critical and a good deal more detailed: the Leopold Report ran a scant 28 typewritten pages, while the Cain Report totaled 207 printed pages.

It made many of the same accusations as its predecessor, with the added charge that little if anything had been done to put into practice the recommendations of the Leopold Report. "Some of the earlier indiscriminate and unnecessary killing of wildlife was reduced," the Cain Report found (large predators known to have been killed in 1970 came to 73,093 coyotes, 8,403 bobcats, 403 bears, 121 mountain lions, and 11 lobos, or gray wolves) "and more selective control methods were adopted." But "as reexamined in 1971, it is clear that the basic machinery of the federal cooperative-supervised program contains a high degree of built-in resistance to change."

This the report blamed on the fact that the program continued to be financed in part by funds from livestock growers; on persistence of "the traditional feeling . . . that the predator control program exists for service to livestock interests" and concomitant lack of adequate recognition of "the growing interest of the general public in all wild animals including predators"; and on "the fact that several hundred control agents today are the same persons for whom for many years the job requirements and measurement of an agent's success have been the killing of large numbers of predators and of personal, uncritical response to the complaints of stockmen. Agents are frequently long-time acquaintances, friends, and neighbors of the individuals demanding service."

The Leopold Report had lamented the lack of the economic research necessary to the establishment of specific criteria to govern decisions on predator control. The Cain Report declared: "A careful review of all cost and damage data available to this committee reveals that no progress has been made to rectify this situation during the seven years that have elapsed since the Leopold Report. Control decisions are still based on the assumption of benefit rather than on proof of need."

A minority of sheep ranchers, the Cain Committee found, did suffer serious losses to predators. It also noted that "no coyote food-habits study has ever shown livestock to be a major part of the diet. . . . Livestock predation, over an entire population, is an infrequent event. It also seems likely that livestock predation is not an activity engaged in by all coyotes." Such scattered data as the committee was able to collect and analyze raised "some real question about the true magnitude of sheep losses and the effectiveness of control programs in reducing total losses. Predator losses may in fact be of such a low magnitude as to be a minor part of total losses."

"Guidelines and good intentions will no longer suffice," the Cain Committee concluded. "The federal-state predator-control program must be effectively changed. It must take full account of the whole spectrum of public interests and values, not only in predators but in all wildlife. This will require substantial, even drastic changes in control personnel and control methods, supported by new legislation, administrative changes, and methods of financing."

Just how drastic the changes were that the Committee viewed as necessary became clear in its recommendations:

1. We recommend that federal-state cooperation in predator control be continued, and that all funds in its support

come from appropriations by Congress and by the legislatures. [This would put an end to participation by the industry in predator-control financing and decision making.]

2. We recommend that immediate Congressional action be sought to remove all existing toxic chemicals from registration and use for operational predator control. We further recommend that these restrictions extend to those toxicants used in field rodent control whose action is characterized by the secondary poisoning of scavengers. Pending, and in addition to, such Congressional action, we recommend that the Secretary of the Interior disallow use of the aforementioned chemicals in the federal operational program of predator and rodent control, and that this ruling be made a standard in cooperative agreements with the states. Moreover, we recommend that the individual states pass legislation to ban the use of toxicants in predator control.

3. We recommend that the field force of the Division of Wildlife Services be professionalized to emphasize employment of qualified wildlife biologists capable of administering and demonstrating a broadly based program of predator management.

4. We recommend that in all states a cooperative trapper-trainer extension program be established as a means of aiding landowners in the minimum necessary control of predators on private land.

5. We recommend that Congress provide some means of alleviating the economic burden of livestock producers who experience heavy losses by predators. [Specifically, Congress was urged to consider the feasibility of some form of livestock insurance.]

6. We recommend that grazing permits and leases written by federal land management agencies provide for possible suspension or revocation of grazing privileges if regulations governing predator control are violated.

7. We recommend that all methods of predator control be prohibited on statutory Wilderness Areas.

8. We recommend that federal and state legislation be passed that would make the shooting from aircraft of wildlife, including predators and game animals, illegal except under exceptional circumstances and then only by authorized wildlife biologists of the appropriate federal and state agencies. [At this point a footnote in the printed report stated that, "Since this recommendation was written the House and

Senate have passed the bill, introduced by Representative John Dingell, that accomplishes the same purpose."]

9. We recommend to the Federal Aviation Authority that a provision be made for suspending or revoking the license of a private pilot and the confiscation of the aircraft—when he knowingly carries a passenger whose acts lead to conviction for illegal predator control, such as shooting from the aircraft or distributing poisons.

10. We recommend that action be taken by Congress to rule out the broadcast of toxicants for the control of rodents, rabbits, and other vertebrate pests on federal lands, and that the possibility of correlative action be explored for private lands as well.

11. We recommend a long-term research program based in the Division of Wildlife Research, Bureau of Sport Fisheries and Wildlife, that would cover the gamut of ecological problems associated with predators.

12. We recommend that the Division of Wildlife Research of the Bureau of Sport Fisheries and Wildlife undertake a detailed socio-economic study of cost-benefit ratios of predator control as a means of evaluating the need for and efficacy of the program and its separate parts.

13. We recommend that the Division of Wildlife Research of the Bureau of Sport Fisheries and Wildlife be delegated the responsibility for the study of the epidemiology of rabies in the field by a team of specialists provided with adequate funding.

14. We recommend that Congress give the Secretary of Interior authority to take measures necessary to protect all species of predators that have been placed on the Endangered Species List by the Federal Government.

15. We recommend that the several states take measures to supplement the federal protection of rare and endangered species by enacting laws and taking measures to protect locally rare populations.

Unlike the Leopold study, the Cain Report resulted in almost immediate and spectacular action. On February 8, 1972, then President Nixon issued Executive Order #11643 banning the field use on federal lands "of any chemical toxicant for the purpose of killing a predatory mammal or bird" and "of any chemical toxicant which causes any secondary poisoning effect for the purpose of killing mammals, birds, or reptiles." And a month later, William D.

Ruckelshaus, administrator of the Environmental Protection Agency, issued orders canceling the registration and prohibiting the interstate shipment of all products containing sodium mono-fluoroacetate, sodium cyanide, and strychnine for use against mammalian predators, and of all products containing thallium sulfate for any use whatever.

Hope Ryden's description of the sheep industry's reaction to these orders verges on the sardonic. "Having for so long believed themselves astride a tiger, the idea of having to dismount filled them with indignation and terror, and they quickly launched a counteroffensive. . . . In January of 1973, just seven months after the poison was presumably removed from the public lands, the National Woolgrowers Association staged a convention in Washington, D.C., at which its members were highly vocal. Not only did they demand that the Presidential Order be rescinded, but reporters at news conferences were given graphic accounts of depredations caused by the 'burgeoning coyote population' which allegedly had resulted from the ban on poison. Journalists who wrote these stories were not told that *Canis latrans* has a single annual breeding season, which had not yet occurred since the removal of the poison baits."

For the time being at least, the days of poisons were ended. Not so the whines: they only grew in volume.

Letters from the Range

Dere Franswa

We had one awful time trackin you down at Jackson, but we done it, as you kin see.

Reason for it is because we have long had relations with ciyotes which we thought you might be interested in.

You might not beleave it but when my sister run off with a travelin man from Iowa and come home with this cute little bastard named Joe a year later which Paw made her to float down the creek in a gunny sack soaked in hog lard, and two ciyotes come down to the bend just below Simons place and they fished that sack out and hauled that young un up into the bresh and when he'd growd he started St. Joseph, down in Mo.

Nother time we was runnin sheep out east a Missoula in a real drouty summer when we ranged onto a old homestead with the well pump still a standing in the yard, but the hole place was long abandoned. The sheep was droppin from thirst and we couldn't get the dam leathers in the pump soaked up soes theyd pump. This dog ciyote come into camp just at sun down and was a sniffin about where our dogs had peed at the well shaft. He lifted his leg and peed down that well, the leathers took up the moysture and we saved the hole sheebang.

Course the next summer the sons of bitches took and et over 200 head spring lambs but we woodnt have had any without them.

We figures ciyotes is kind of a mixed blessing. Hoping you are the same.

Sincerely,
Alfred Potage
Shorthair, Utah

This was but one of nearly two hundred letters I received during the first half of 1975 in reply to a letter to the editor I placed in newspapers and magazines across the nation, in which I asked for personal experiences with and opinions about predatory animals. Letters poured in from sheepmen, cattlemen, hunters, trappers, wildlife biologists, conservationists, and a wide assortment of men and women with an interest in, and either a liking or a hatred for, predators. Of course I have reservations about Mr. Potage's letter. For one thing, it was addressed to me in Jackson, Wyoming, where I happened to be spending three weeks on a *National Geographic* assignment. How could he have known I was there? For another, a careful perusal of the atlas failed to reveal the location of Shorthair, Utah; and the letter, in fact, was postmarked San Rafael, California. Finally, the handwriting—or handprinting—bore suspicious resemblance to that of Osborn Howes, my good friend and next door neighbor in Kentfield, California, a schoolteacher and part owner of a cattle ranch in Idaho! But whatever its authorship, the letter seems worth quoting for its fine, balanced appreciation of the coyote's pros and cons.

Another letter, the authenticity of which might be questioned, is worth including for comic relief. It came from Silver City, Nevada:

Dear Mr. Leydet,

They showed me your letter in the paper today. I got a story about a coyote.

I am 97 years of age, I will be 98 on March 23. I have lived up here on the Comstock all my life. I used to have a friend named Ambrose Evergreen. He lived in a little shanty up behind what usta be the May Day Mine. Later he came to a bad end, but that was later.

Back in 1906 Ambrose Evergreen and I went down to San Francisco. That was in the spring & we didnot get back till the fall.

When we got back the news was they'd had a hanging down to Dayton & they'd hanged the wrong man, & up in Silver City there was a coyote had been eating Old Man Ross's chickens. Old Ross was the old man who ran the Golden Gate Bar & he was a mean old cuss, & he was plenty mad that his chickens was gettin et. So he offered a bottle of licker and, oh yes, I forgot to tell you, he also had a cat house up above his bar, so he also offered a night with one of his girls for free to the man who could get the varmints that was eating his chickens.

Well it was a pretty mean crowd of miners that was up here in those days & they tried to get the varmint. They usta set out by the hen house with there guns & wait.

Well, the truth was there weren't no varmints at all stealing them chickens. What Old Ross did was sell them so much licker to drink while they was settin out waitin up for the coyote, & then like as not—it being fall & mighty cold that fall—the men ud be so danged

*cold they would need to sleep with a floozie that night to get warm
again. And Old Ross would sneak out & steal a couple of chickens
& spread some blood & feathers around when everyone was gone,
to make it look like the coyote had come callin. He was cleaning up
all right with his little scheme, but of course didn't any of us know
that.*

*Anyway when I & Ambrose got back to Silver City we went to
stay at his cabin which was still there & not bothered because it
was such an old shanty while my family had taken the opportunity
to move on & leave me & so I didn't have my own place to stay.*

*When Ambrose left he went off and left a great big iron pot of
stew atop his stove. Well when he got back the stew was still on his
stove rite where he left it. Of course it had gone bad. Well the only
thing was to throw it out, but it was mighty cold by the time we had
climbed up the hill back of what uster be the May Day Mine, so
Ambrose just slung the whole kit and caboodle out the door. Well
in the middle of the night we heard a noise and we went & looked
out. There was this great big coyote eatin on the stew & not paying
us no attention. Well we watched & watched & the coyote didn't
pay us no heed most likely I figure because he was makin to much
noise nuzzling round to hear us & also the smell of that old stew
kept him from smelling us, though that was some wonder because
Ambrose warnot the best smelling feller I ever knew, particularly if
you bunked with him because of his socks, which he reckoned it
unhealthy to change to often.*

*Anyway that coyote didn't seem to know we was there. All of a
sudden he sort of gasped and sprang back and his expression, even
for a coyote, was not pleasant to behold. His eyes was bugging out
& his tongue was lolling out & his expression, if you can say a coyote has one, was awful. He still didn't see us but went sashaying off
in a terrible state.*

*The next morning Ambrose & I found him down by the creek. He
was deadern a doornail. So we skinned him out & took him to Old
Man Ross & said "Did any of youre chicks get et last night?" He
looked at us sort of suspicious & said "I aint check yet. Why you
boys asking anyway?"*

*"Well, course they did not," we said because we figured there
sure warnot two such coyotes hanging around Silver City so this
must be the one for sure.*

*But Old Man Ross didnot want to pay us because he knew no
coyote had been eating his chicks and besides he didn't want the
game to end.*

Well it was payday that day & all the men outer the mines pretty soon was feelin pretty good, & besides they was tired of looking for the varmints themselves, so they said to Old Ross, "Give these boys their rightful prize or you know what will happen to you."

Well Old Ross was madder then a wet hen but he didnot want to have his ass painted blue & be rode through town atop the fire engine, which was a custom we had in those days before the county finally built the jail here, so he fetched up. Only he didnot give us but one bottle of rot gut and one girl. Ambrose grabbed off the girl for himself & so I only got the bottle of rotgut. But Ambrose got the clapp so I warnot so mad afterwards.

I got a good story about a deer up here, but I guess that isnot what you want as you state "predators."

I hope you get other letters from people that read the paper. Goodluck on the book. I got to go to the toilet now so I got to close. God bless you & good luck.

Your friend,
Henry Park

There was no doubt, on the other hand, about the authenticity, or sincerity, of the many letters I received from stockmen in a dozen western states.

Mrs. Tom Carey, the rancher's wife in Boulder, Montana, who made the statement quoted in Chapter 1 about "the last coyote on earth," continued in her letter:

I think, at one time perhaps, every rancher in this Valley, stretching between Whitehall & Boulder, has thought to himself, "Maybe coyotes aren't really as thick as we thought—maybe the ecologists are right—maybe they don't kill stock." Every rancher in this Valley today will tell you he's changed his mind. Not only has he lost tiny calves & lambs to coyotes—he has actually witnessed the kill so knows that no other predator was responsible.

I'd like to tell you some of the bloody stories regarding coyotes that we know of in just our small area.

One neighbor saw from his window one day, away up on a side hill, two coyotes working together to get a calf from a cow. One would harass the cow. The other would sneak in and drag the live, healthy calf a few feet away. Then the other would distract the cow and again the calf would be dragged. The man did get up in time to save the calf—bleeding badly.

Another neighbor, Martin Carey, was not so fortunate. Upon seeing a young cow quite agitated, he drove up to find a coyote eating her calf alive. *The calf had been completely bled out by the coyote eating its entrails out the back end. When Mr. Carey put the calf out of its agony—the bullet hole in the head didn't draw a drop of blood & yet the calf was still alive when he found it. . . . The John Careys were much luckier. Their herdsman, while making his rounds one day, happened upon a coyote eating a live calf. The coyote was on the other side of the fence from the cow—the calf partially on his side. He was smart enough to realize the cow couldn't get to him. The herdsman chased the coyote off and brought the calf in. Through excellent care the calf did survive—* tailless *and* earless. . . .

Another Montanan, Harold Anderson of Lavina, assured me that:

. . . the wild organizations do not need to worry about ever exterminating the coyotes. He will be here when you and I are not.

The way a coyote can adapt to changing conditions is uncanny. I met the first government trapper in 1924 before I graduated from high school. The predatory animal program was in operation for 40 or 50 years before the do gooders knew it and we still had coyotes.

This man taught me how to find coyote dens. . . . if you got within a quarter of a mile of the den the old coyote dog would start to bark and howl and attack my farm dog if he got too far away from me. . . . When the coyote started yapping you were close to the den. . . .

Now you can ride with a dog right over the den and they stay absolutely quiet. Not one bark. They must have learned that the barking led you to their den. Dens are really hard to find now. You just accidentally stumble on to them. . . .

A coyote is the most destructive thing God ever put on this earth. . . . It all depends on whose pocket is being eaten on. I have seen some genuine coyote lovers change to coyote haters as soon as the coyotes started killing their calves. . . . I have heard people say they liked to hear them howl. Someday, probably not in my life span, the people in the cities will have to decide if they rather eat or look at a lot of predators and listen to them howl. . . .

All we hear is alternate ways of control and the coyotes are still killing while the law makers and wildlife enthusiasts are making laws & studies. We know what to do but they took the poison all away from us.

Charles Dean, of Fruitland, Idaho, would agree. He wrote,

Many of the heads of state game departments are getting their information from books written by liberals and misguided conservationists. . . . Many of the conservationists get their information from a Walt Disney movie made for six-year-old kids. . . .

We had the near perfect tool for coyote control in the poison 1080. It was selective in that it killed only members of the canine or cat family when it was used at recommended strength. It was used only by members of the U.S. Fish and Wildlife Service who were trained in its use. It was used only in remote areas where coyotes were killing livestock or wild game. . . . Since [1972] there has been a population explosion and many farmers report as high as a 40 percent loss of lambs. . . . The sheep industry is about a thirty-million-dollar industry in Idaho alone and predators are putting many growers out of business. When sheep are gone the predators eat calves and wild game and birds. . . .

[Coyotes] will never be exterminated completely. It is just impossible because they are too intelligent and adapt too well. . . .

Albert Peterson, of Stanton, North Dakota, gives an instance of this adaptability:

. . . aerial hunting . . . is O.K. if the man handling the gun is a good shot; if not, a coyote that has been shot at a couple times gets smart and will run into a hole or other cover and cannot be seen from the air, if he has been shot at several times he will go for a hole as soon as he hears a plane. . . .

In areas where there are sheep, [coyotes] should be completely killed out. . . . two years ago I lost 35 head of sheep to coyotes before they were controlled, which was a loss of at least $1,000 to me. They usually would take one every two of three days early in the morning sometimes one every morning, and then get spells when they would kill just for the fun of it. . . .

It would seem to me the ecologists and environmentalists who wish to preserve the coyote are not thinking very straight, why should a killer coyote be allowed to torture an innocent lamb, eat a small part of it or leave it nearly dead and go for another one, be allowed to roam the land and kill when so many people are starving in the world?

Apparently many sheepmen have not heard, or choose not to believe, the criticisms that have been made of the 1080 program. One who came to its defense was Henry V. Rowe of Norris, Montana:

When the 1080 came out, it was handled by one government trapper who poisoned the baits for the rest of the government trappers in a large area and was properly handled. . . . In the coyote years, if we survived with an 80 percent lambing at shipping time, we thot ourselves quite lucky. After the 1080, many falls we shipped 118 percent and the weight went from a 75-pound lamb to some 90-pound average. . . .

I live in the Madison Valley, one of the most beautiful stock rais-ing areas in this state and, I believe I am safe to say, in the West. At one time, this valley was assessed for over 60,000 head of sheep. I doubt that you could find 1200 in all now, counting the farm flocks and all. Coyotes can be seen in numbers now all around this area. In the spring, they are killing small calves and are beginning to be a problem to the cattle industry. . . . There is a lot of good ranches in this area that are now operated as a cattle ranch and are loosing out because it is a better sheep ranch, but what else can they do?

John McRae, a young rancher of Miles City, Montana, expressed more bluntly than any other correspondent the attitude that the earth was created for the benefit of man and that man is justified in using any means to increase his dominion over it:

A controlled program to control predators is the best way I can imagine to have our natural environment adjusted by man to inhance his livelihood. The controls should not be just for pred-ators, but for any effective means to inhance our subsistence. . . .

The preponderance of letters I received from stockgrowers— certainly from the sheepmen—were strongly anticoyote. Perhaps those men who were experiencing losses felt more moved to write. But there were others, too, whose attitudes towards *Canis latrans* ranged from neutral to friendly. Among the neutral was A. P. Atkins, president of Atkins Ranches, Inc., of Guymon, Oklahoma. He wrote:

This is open plains cattle country—sheep population negligible. I have never experienced any appreciable calf losses to coyotes. I would rather have coyotes than stray dogs, who sometimes hunt in packs, and I have had considerable damage to fences, and gates left open by hunters with packs of hounds. We do not permit coyote hunting on our ranches.

However, I feel that an uncontrolled coyote population could result in consequences not foreseen by the ignorant environmentalists who want to abolish all predator control. The natural forces which maintained the primitive ecology of America have been destroyed by urban civilization and no sentimental nostalgia can bring them back. Perpetuation of endangered species is the proper function of zoos and wildlife refuges. . . .

The most practical and humane measures to control coyotes are cyanide guns and poisons, which kill quickly, and den hunting. Traps are inhumane, ground and aerial hunting are too expensive, and bounties are subject to cheating. Coyotes are too smart to be eliminated entirely, but voluntary control can keep their numbers within reasonable limits.

Another rancher who, like my friend Alfred Potage, seems to view the coyote as a mixed blessing is W. B. House of Cody, Wyoming.

My information comes from living in Wyoming since 1915, and having experience in cattle and sheep raising, farming, and traveling over practically every section of this state, as well as Montana, Colorado and other mountain states.

The predator problem—like most any other problem getting a lot of publication—is not serious enough to deserve much attention; or anything like it has. It can be a problem for the sheep raisers if certain conditions exist, tho nothing like as serious as they claim.

I owned and herded sheep for two or three years way back in 1918, and in that period I only lost one lamb from coyotes. However, I did a good job of herding, being close to the band of 1700 ewes, with lambs. The one that the coyote got was when I was finishing the breakfast dishes at the sheep wagon, and the head of the band got over a hill where I couldn't see it in time. The coyote bit him on the neck and back of the head, and when one bites it's all day with the lamb. I never had a coyote get a lamb or ewe from the band when they were on the bed ground. . . .

Where big lamb losses are claimed, or in existence, the main trouble is that they are not having proper attention by the herders. They are lazy, and don't care what happens to their flocks. It's hard to get good herders, but when good ones are with the herds the loss is very nominal, and only lasts a few weeks when the lambs are quite young. Coyotes bother large sheep very little, tho occasionally a bear up in the mountain areas will kill several in a band. . . .

The coyote is extremely smart and wary, but he isn't worth a darn for anything except to help keep the balance of nature. If he gets too numerous he will, thru hunger, eat lambs or anything else he can catch. If he gets killed off entirely, rabbits and other things will be as bad or worse. . . .

Clifford McElrath, of Jackson, California, retired from ranching and appraising ranch property, wrote in defense of the coyote.

The coyote is much maligned. I have been out with cattlemen who should know better and on coming upon the remains of a calf would immediately commence to cuss the goddamn coyotes. On more than one occasion I have gotten off my horse and showed them the tracks of a mountain lion who did the killing. The coyotes had been feeding on the lion's kill. I have discussed this with other old cattlemen and they all agreed that they had never seen a case of a coyote killing a calf.

A cow will leave her calf hidden in the grass while she goes to water for about the first two or three days after it is born. A calf at that age is like a young deer. He has no odor for the first few days and a coyote will pass very close without noticing him. What the coyotes are looking for is the afterbirth. Added to cleaning up afterbirths they kill a lot of rodents.

Sunday hunters will never miss a chance to take a shot at a coyote and I know ranchers who carry rifles in their pickups to shoot coyotes. They do not know that coyotes and quail particularly have a close connection. Get rid of your coyotes and you can say goodbye to the quail. I have seen this demonstrated very clearly at least twice.

An old-time sheep herder from Boulder, Colorado, Ernest Ross, writes that

. . . in my younger days I have put in a lot of time when a coyote howl was the only company I had for a month at a time. I sure hate

to see them all killed. I can say a lot for them. I have heard people tell how a coyote kills sheep just to kill. Not so. I have herded sheep and have seen where a coyote had killed one to eat. It is always an old or sick one that can't get away very fast, never a young healthy one.

Then I have had dogs get in a herd and kill just for killing. One night I remember a dog killed 25 sheep but never ate a one, not a mouthful even.

William Glaser, of Deeth, Nevada, writes that he has lived all his life on a ranch in Elko County, and that

. . . since Elko County is one of the largest cattle and sheep areas in the United States I believe we have one of the largest coyote populations also. . . .

One year I lost 9 head of large lambs to an old bitch that had pups. She made a visit to my herd of 60 sheep once a week on the same day at about the same time of the day but only took one lamb each time. This is why I was able to get a shot at her. However it didn't kill her but did cripple her for the rest of her life and I seen her for years after as she hunted the fields for mice every summer.

Very seldom do you see coyotes attacking livestock in large packs. I have only witnessed on one occasion seven coyotes pulling down one ewe. If this was the case packs of coyotes could kill large bands of sheep in a short time.

About the only time coyotes run in packs is during mating season and I have counted as many as fifteen in one bunch. This is usually during the month of February and very seldom do they fight among themselves, however they will take after a dog and sometimes chase them right under the wagon, tractor or truck.

Sheepmen blame the coyote and paint a terrible picture of how this predator kills his sheep, but I have seen small bunches of sheep in the mountains that have strayed away from the main band and wandered around the mountain side for months and finally die do to winter weather, the coyotes could make short work of these strays if he was as vicious as he is supposed to be. However, the predators do get a certain per cent of lambs and older sheep.

Coyotes do attack and kill small calves but this is very seldom and only when he becomes very hungry. I know of only three calves that I have lost over the years to the coyotes and when they do make a kill as a rule it is a male and female and they work as a team, to do this one always attracts the cow's attention and draws her away from the calf while the other one makes the kill or they

might lay around for hours untill the cow goes to water and then attack the calf. However most of the time a number of cows will stay together and some will always stay with the calves while the others go off to water or feed.

Coyotes do fish as I watched one several years ago that was fishing a small stream and would jump into the water, grab a trout then place it upon the bank and go after another. After a while I run him off and took the fish home. I will never know if he was doing this for fun or food.

I am very much against the use of poison as this is a terrible death and very dangerous to use. I have seen many beautiful birds, dogs, cats and even livestock die from this type predator control.

I am against extermination of any animal or bird, but the so called predators should be kept under control and thinned out when necessary. Some people think the coyote should be exterminated completely but there is definitely a need for this animal.

The coyote like the magpie and vultures are very beneficial to the rancher at times as they will eat all carrion or dead animals that is left around besides all rodents.

In my opinion the best way to control the predator is through bounties executed by all western states and handled by the state wildlife management, each state paying the same amount of bountie money and by trapping or ground shooting only.

Finally, I have a letter from one old cowboy who actually formed a friendship with a coyote while punching cows for an outfit in Arizona. Bill Giles, of Crow Agency, Montana, states that "Letter writing just aint my thing," but he does a pretty fair job of it just the same:

I will be 66 the later part of this month and have been watching coyotes for the past 40 years. I have been well acquainted with two government trappers and have known many old time sheep herders. For the past 25 years or so, I have been what you might call a saddle tramp. I have punched cows throughout the Rocky Mt. region—almost from border to border. I have never known for sure (positively) that a coyote ever killed a calf. If a coyote ever killed a calf, he wouldn't quit killing. He wouldn't kill one or two calves and quit. He would go on killing calves until someone killed him.

I have never seen a range cow that paid any attention to a coyote —they don't pay any more attention to a coyote than they do to a jack rabbit. But when a dog shows up they are on the alert right now and start running to their calves. That is because at some time or other they have been worked with dogs, and fear a dog. If a coyote ever caught a calf, the calf would surely let out a squall and every cow within hearing distance would come on the run. The cows would soon learn to fear a coyote as they do a dog. They sure know the difference between a coyote and a dog.

I had a camp job when I was working for Camden Livestock Co., Seligman, Arizona, and made friends with an old coyote. It took almost two years for me to gain his confidence but finally he came to consider me as being more of a friend than a foe. I would save up my table scraps and when I had enough to make a meal for him, I would whistle and pretty soon I would hear him answer. He would bark a few little short barks then after a little while he would show up. In the winter time, he would stay around camp, then in the spring he would take a mate and I might not see him for a week or ten days. When he stayed around camp he never bothered anything. There was a covey of quail that wintered at my camp and he never bothered them. My old tom cat was pretty wise and knew how to take care of himself. I don't know if the coyote ever took after him or not. I got him to follow me twice when I left camp horseback. He would follow for about a half or ¾ of a mile then he would swing around and head back towards camp. . . . That old coyote gave me a lot of good laughs and it would take me a long, long time to put it all in writing. He was the best watchdog I ever had. Sometimes, early in the morning, I would hear him bark a few little short yips and in fifteen or twenty minutes the ranch pickup or truck would show up. The boss tried, but he never caught me sleeping in.

From what I have seen at first hand and from what sheep herders and government trappers have told me, I am convinced that only a very small percentage of coyotes kill sheep. But once they start killing and find out that that is an easy way to make a living they keep right on killing. Frank Morris who was the government trapper in the Cody, Wyoming area and the government trapper at Absarokee, Montana, both told me that when a coyote starts working on a band of sheep, you can't poison that coyote and you can't trap him. The only way you can get him, if you can do it, is to out figure him and get a surprise shot at him. Sometimes that takes quite a while. No doubt where sheep are turned loose in a large pasture without a herder the coyotes will kill quite a lot of them. Sheep have been domesticated too long to know how to take care of themselves. Frank Morris told me that the coyotes in his area had gotten wise to cyanide guns and that it was a waste of time to put them out.

From what I have seen with my own eyes, I am convinced that coyotes do a lot more good than they do harm. . . . They get blamed for a lot of things that they don't do! . . .

7

The Numbers Game

WHEN I RE-SCAN all the letters I received, only a fraction of which, unfortunately, have I had space to include, the primary impression I draw is of an industry in trouble. A few of the writers acknowledge problems other than predation—labor, inflation, wool prices, the weather, grazing restrictions on public lands—but they do so only in passing. It is as if all these difficulties are facts of life to which they are resigned and about which they feel they can do little or nothing. But the coyote is a different matter. "We *know* what to do about *him*," many of the ranchers say. "But our hands are tied thanks to meddling environmentalists who know nothing about ranching. Just give us back the 1080 and we'll take care of the coyote; the other problems will take care of themselves or at least will not be so great that we cannot live with them." Still, there *are* voices of dissent from this majority opinion.

If there is a sub-pattern to be found in the letters, it is that sheepmen are much more virulent than cattlemen in their hostility to the coyote. But even this pattern is not consistent, and I am left perplexed. Whom am I to believe—Mr. Rowe, who claims that environmentalists and coyotes are putting "the sheep industry out of business," or Mr. House, who writes that although predation "can be a problem for the sheep raisers if certain conditions exist," it is "nothing like as serious as they claim"? Mrs. Carey, who bewails "the horrible maiming and killing the coyotes do here and the agony of the livestock industry over its extensive losses," or Mr. Glaser, who states that "coyotes do attack and kill small calves but this is very seldom and only when he becomes very hungry," and Mr. Giles, who befriended a coyote in Arizona and who has "never seen a range cow that paid any attention to a coyote"? Or should I believe them all?

My inclination is to credit statements of particulars, such as Mrs. Carey's description of actual coyote kills or Mr. Glaser's account of seeing bunches of stray sheep wandering around the mountainside, but to remain skeptical of generalized conclusions either that coyotes never take calves or that they are driving the sheep industry over the cliff. I tend to believe Dick Randall's statement that predation "is a very local thing. [It] occurs in one place and over the hill it doesn't occur, and some sheepmen have a lot of predation, others have very little. Part of it is due to the area they operate in, part of it is due to management of the livestock." On the day following my meeting with Dick I stopped in Rawlins and had a talk with Dr. James R. Tigner of the Fish and Wildlife Service, who had been conducting a study to assess and identify sheep losses associated with range lambing and herded grazing operations. His words were much the same as Dick's:

"Overall, and based on total number of animals, we've found predation running about 5 percent. But this figure is not accurate, or at least it's misleading. It really masks the problem, which is that you can have significant predation in certain areas, at certain seasons, and little or none at other times, other places. Some areas historically have more predation than others, but even this varies from year to year. Management? Sure it plays a part. One operator can come off the forest and count his missing animals on one hand. Another may be missing a hundred or more."

But why, aside from differences in quality of management, do predation losses vary so much from one area to another? Why are some coyotes livestock killers, others not? What percentage of the coyotes in stock-raising areas turn to killing calves and lambs? The truth seems to be that nobody knows.

"New Mexico is a classic example of the numbers game bullshit," Brant Calkin warned me on my first day in Santa Fe. Then president of the Frontera del Norte Foundation and now president of the Sierra Club, Brant showed me a clipping from the March 23, 1974, Albuquerque *Tribune. "Coyotes prowl N.M. in record numbers,"* the headline said, and the article went on to cite a 53.59 percent increase, in one year, in the state's coyote population.

Since nobody in New Mexico knew if there were 50,000, 100,000, or 200,000 coyotes in the state, the precision of the alleged 53.59 percent increase in the coyote population was surprising. The article mentioned the figure only once, attributing it to the U.S. Fish and Wildlife Service but giving no particulars on how the percentage was computed.

"That figure," Brant told me, "was taken from Fish and Wildlife's transect line study and represents the change between 1972 and 1973 in the statewide coyote visitation index. But it is a gross misinterpretation to claim that the population increased by that amount." And indeed it was, later agreed Bob Roughton, who had helped develop the comparative predator abundance indexes and with whom I ran the scent post lines in February.

"An increase or decrease in the index from one year to the next," he explained, "measures an increase or decrease in the number of visits by certain species to our scent posts. A substantial rise, like the 54 percent in New Mexico between 1972 and 1973, combined with the high probability factor of 0.99 for that year, indicates that most likely there *was* a rise in population, but it need not have been anything like 54 percent. All kinds of factors other than population affect the rate of visitation—the weather, the availability of prey, and so forth. Our index survey methods are still at the developmental stage and we try to discourage anyone from trying to jump to conclusions from these data."

Another instance of the "numbers game b.s.," Brant Calkin said, was the 1971 report on livestock losses by Dr. Jack L. Ruttle of New Mexico State University. A questionnaire was mailed to 2,500 livestock producers in the state, asking them to enumerate their losses during the previous twelve months from four categories: diseases, poisonous plants, predators, and other causes. Responses came from 460 individual producers, who owned 15 percent of the cattle and 26 percent of the sheep in the state. Ruttle took the claimed predator losses (10,213 lambs, 2,742 ewes, 823 calves, 24 cows—or 60.4 percent of total losses from all causes of 22,842 animals), multiplied the sheep losses by 3.8 and the cattle losses by 6.6 to obtain a state estimate, multiplied this by a 3.0 "income multiplier," and announced that economic losses to New Mexico due to predators totalled $5,479,375.95, a figure that was promptly proclaimed by the press.

"What kind of idiotic statistics are those?" Brant asked. "Some of the ranchers who didn't respond may have had such losses that they were driven out of business. But the big majority, I'll bet, didn't answer the questionnaire because their losses weren't sufficient to bother reporting. If this was the case, Ruttle's total was a gross exaggeration. But it's typical of the numbers game when it comes to coyotes and sheep."

My preference is to steer clear of statistics, but I could not adequately describe the status of the coyote without delving into some data. The picture remains incomplete—my frustrating search for reliable statistics stopped after Arizona and New Mexico—but does provide an overview of estimates in three areas: the coyote population; the number of coyotes killed each year by man; and the numbers of sheep and cattle killed each year by coyotes.

The New Mexico Department of Game and Fish had devised a somewhat hit-or-miss formula to estimate the coyote population and had arrived at a total of 144,749 coyotes in 1975, up 15 percent from the previous year. Divided by the state's 122,666 square miles, this works out to a density of a little more than one coyote per square mile. I did not find any authorities, outside the Department of Game and Fish, who regarded this coyote census as very scientific or reliable, but I also did not find anyone who could offer proof that the total was too high or too low.

Five years ago, in Arizona, Wildlife Services director Robert Shiver numbered the coyote population at 115,000—or one per square mile. Conservationists then called the estimate "nonsense," and now William Rightmire, who as Fish and Wildlife Service state supervisor for Animal Damage Control performs Shiver's former duties, would not even hazard a guess as to how many coyotes there are in the state. Nor would Robert A. Jantzen, director of the Arizona Game and Fish Department. Nor would any of the university wildlife biologists with whom I spoke. Dr. Roger Hungerford, at the University of Arizona, thought that coyotes might be "that thick"—one per square mile—in some areas, but that the figure would be "way too high" for vast expanses of creosote bush desert where prey populations would themselves be very sparse.

I did discover virtual unanimity of opinion on one point: in neither Arizona nor New Mexico is the coyote considered an endangered species. In fact he is prospering. General consensus says that coyote numbers are increasing in most areas, and the only issues in dispute are the degree and the cause of this increase. The sheepmen in particular see the ban on toxicants, especially 1080, as the cause for the rise in coyote numbers. Some wildlife specialists view the increase as part of a natural cycle, governed more by the availability of natural prey than by the absence of poisons from the coyote's domain.

What about my second statistical quarry: the number of coyotes killed annually by man? Here at least there were some fairly accurate figures from 1972 on—Animal Damage Control's (ADC's) annual count of animals taken. Before the ban on toxicants it was impossible to know with any certainty how many animals had died from 1080 or strychnine poisoning, since an animal might travel far from the bait station before succumbing and its body might never be recovered to be counted. Table 1, in Appendix B, shows the number of coyotes taken from 1964 through 1975, in Arizona and New Mexico. (Tables 2, 3, and 4 give a more complete breakdown of the "take" in 1975, showing other species killed and the method of killing.)

But the ADC figures tell only part of the story, and a small part at that. As Donald Balser said to me in Denver, "We are killing every year at least five times as many coyotes by all-around human pressure as the federal cooperative control program takes. ADC took 74,000 coyotes in 1974. We know of 224,000 that were taken by others besides our trappers. Probably a half million or more coyotes are killed every year by man. And still he survives."

How the coyote manages to absorb such losses was suggested in a 1975 computer study, *The Effects of Control on Coyote Populations*, by Guy E. Connolly and William M. Longhurst of the University of California at Davis. "According to a model developed to simulate coyote population dynamics," the authors state, "the primary effect of killing coyotes is to reduce the density of the population thereby stimulating density-dependent changes in birth and natural mortality rates.

"Tests of varying levels of control kills showed that a coyote population can maintain itself and even increase its numbers except at the very highest levels of control. If 75 percent of the coyotes are killed each year, the population can be exterminated in slightly over 50 years."

Of course this conclusion, Dr. Connolly cautioned me in a letter,

. . . is not based upon field study but rather upon a theoretical, mathematical model [that] uses untested, graphical hypotheses which may or may not be correct, but which certainly are oversimplified from the real world. I am not willing to have this statement applied literally to every coyote population in the West. However, there is no doubt that coyote populations can withstand an amazing amount of control.

There are those who do their best to reach or better this kill figure of 75 percent. "A lot of landholders shoot any coyote they see," New Mexico Game and Fish Director William S. Huey said. Of course shooting and hitting are not necessarily synonymous, as Harold Anderson made clear in his letter, part of which I quoted earlier: "I have shot [coyotes]—not too many either, as they do not give you many good shots, and they are a lot of fur and not more than six inches of carcass at the largest part. . . ." Still, a good many coyotes *are* shot and killed by ranchers, and trapped and poisoned, too—there is a black market in toxicants, I was reliably informed, including 1080 smuggled in from Mexico, and poisons although now illegal are by no means absent from the range. But there is no record at all of how many are killed this way.

One of the most common sights on roads and streets of the West, at least outside the big cities, is the pickup truck with a rifle rack mounted across the rear window of the cab. Many are adorned with bumper strips warning that "When Guns Are Outlawed, Only Outlaws Will Have Guns"; one even proclaimed, rather frighteningly, that "God, Guts and Guns Made This Country Great"! I commented on this to Dr. Gary R. Olsen, assistant professor of history at the New Mexico Institute of Mining and Technology at Socorro. "Many of those vehicles' owners are no more ranchers than you or I," he said. "But it's a status symbol around here. It's part of the he-man *macho* ethic to drive a pickup and carry a rifle." Status symbol or not, the guns are made to shoot. And no doubt many coyotes are shot, not only as "pests" by ranchers, as competitors for game by deer hunters, as game themselves by varmint callers, but as targets of opportunity by grocery clerks and gas station attendants driving around in their free time looking for trophies to prove their *machismo*. But of course there is no tally kept on these kills.

Nor is there any tally kept on the number of coyotes trapped for fur, either by professionals or by weekend trappers. But with fur prices currently high, the number is known to be large. (Several Game and Fish men, in both New Mexico and Arizona, expressed concern to me about the impact trapping might be having on bobcat populations. With prime pelts fetching as much as $200, bobcats, far less numerous than the coyote, much less prolific, and much easier to trap, could be in trouble.) To obtain some idea of the number of coyotes being trapped, I paid a call on the Cox Fur and Hide Co. in Albuquerque, New Mexico's largest dealer in pelts. I bumped into R. L. Cox just as he was leaving his establishment, and I explained the purpose of my visit. He looked at me suspiciously, told me that neither he nor any of his employees had the time to talk to me just then, that he was working till eleven o'clock every night to process the deluge of furs pouring in.

"Come back in a month," he suggested. "Perhaps we'll have time to talk then. This has been the busiest season we've ever had." Unfortunately I could not come back in a month, and the volume of the fur trade in New Mexico and Arizona remains unknown to me, as it apparently was to all the experts I consulted.

Still, by making some assumptions, and juggling a few figures, we can draw some (highly tentative) conclusions. Assuming former Wildlife Service state director Robert Shiver's estimate of Arizona's coyote population, and the Department of Game and Fish estimate in New Mexico, are 50 percent high, then rounding out the figures, we could say that Arizona harbors 76,000 coyotes, New Mexico, 97,000. Let us assume further that Donald Balser's guess—that all-around human pressure kills five times as many coyotes in a year as are taken by ADC—is reasonable and holds true in these two states. Multiplying ADC's 1975 coyote take by six, (or their take plus "human pressure's" five times that), we could calculate that the total man-caused coyote mortality for that year was 8,000 in Arizona, 31,200 in New Mexico. This would represent approximately 10 percent of the base population in Arizona, and just less than one-third of that in New Mexico—or well below the 75 percent annual kill which the computer study predicted would have to be achieved for fifty consecutive years to eliminate the coyote. Assuming, finally, that the assumptions Connolly and Longhurst made in programming their computer were valid, we might conclude that man has little or no effect on coyote populations in the two states, that William Rightmire was essentially correct when he said to me that "coyotes in Arizona are controlled by natural factors," and that the man-caused mortality in coyotes in Arizona and New Mexico is substitutive for, rather than additive to, natural mortality. We might even be tempted to wonder whether, at its present level of intensity, coyote control does any good at all, unless it is very precisely targeted at individual livestock-killing animals. If intended as a population repressant it might be a waste of effort and money, for we could be merely skimming off the population surplus, killing coyotes that, if left alone, would succumb anyway to natural mortality agents.

What is important, however, is to remember that in killing these thousands of animals without being very sure of what we are accomplishing, we are playing around with lives, which involves questions of ethics. And we certainly are playing around with complex ecological systems with very little understanding of the rules of the game, knowledge of the odds, or appreciation of the stakes involved. Positing an admittedly extreme, and I think most unlikely, situation, I asked Dr. Michael L. Rosenzweig at the University of Arizona what might be the consequences if we were to succeed in exterminating the coyote in the Southwest. Dr. Rosenzweig, one of the world's leading population ecologists, smiled and answered: "We can't say for sure, but owing to the effect this could have on the rodents, we could ruin the vegetative cover in the Southwest and we'd just flush down the Colorado River!"

If I was more or less stymied in my search for reasonably accurate estimates of coyote populations and of the numbers of coyotes killed by man, I had no better luck in my quest for dependable figures on the number of livestock killed by coyotes. The "paucity" of such figures appalled the Leopold and Cain committee members. If any other industry were to suspect, or know, that some factor was consistently causing it substantial losses, no doubt considerable effort would be made to identify the offending factor and the exact amount of loss attributable to it. But not so the livestock industry, and of course there are special problems involved in locating dead animals out on the range, and soon enough to prevent scavengers and decomposition from obliterating evidence of the actual cause of death.

Only 55 woolgrowers, out of 300, responded to a questionnaire sent out by the New Mexico Department of Agriculture. One-third of these responses were too incomplete for tabulation, but those that were complete claimed an annually increasing loss of lambs to coyotes, from 1.6 percent in 1970 to 7.7 percent in 1974. But why the apathetic response?

In Arizona, the Woolgrowers told me that they had passed on their loss tabulations to the Cattle Growers, but John Olson, Executive Vice President of the Arizona Cattle Growers Association, informed me that a thorough check of his association's files had failed to turn up these reports. He did furnish me, however, a tabulation of losses suffered in 1974-1975 by range cattle operations in his state, some of which also run sheep and other livestock (Table 5 in Appendix B).

But here again, although the response was better than in the New Mexico survey, only 58 percent of the members polled returned the questionnaire (nearly half of them reported no predation losses). Why not the other 42 percent? Were their losses negligible? Were they too busy to answer? Or did they have a fatalistic attitude that whatever information they returned would be filed and forgotten and lead to no action?

The U.S. Department of Agriculture's Statistical Reporting Service published figures for 1974 showing that 9.2 percent of docked lambs were killed by coyotes that year in New Mexico, and 6 percent in Arizona. "Other causes" of loss were 8.1 percent and 2.3 percent, respectively. S.R.S. statistics were compiled from reports from about 9,000 producers surveyed in 15 Western states. Individual producers differ in the accuracy of their records, as they do in the degree of their bias towards coyotes. And such imponderables no doubt affect the reliability of the data.

"We like to think all our members are honest," Olson said to me. "But we can't absolutely guarantee their figures." At New Mexico State University, in Las Cruces, Dr. V. W. Howard, co-author of a sheep predation loss study, told me it was his belief that in New Mexico "the average rancher has been giving a close estimate, perhaps even an underestimate, of his predation losses." On the other hand Dr. William P. Stephens, director of the state's Department of Agriculture, thought "most ranchers can't identify their losses—at least they find it very difficult to do."

It may be that all these computations are academic in any case. For one thing, it is a common assumption, in assessing the economic costs of predation, that animals killed by predators are a net loss to the economy. Of course that is not so. Just as man-caused mortality in a coyote population is at least to some extent substitutive for natural mortality, even so an unknown but perhaps considerable percentage of lambs and calves killed by coyotes and other predators would, in the absence of the predators, have died anyway of other causes before ever reaching the market.

Furthermore, as Dick Randall and Dr. Tigner made clear, overall loss figures are essentially meaningless. ADC's Terry Anderson put it succinctly: "We have to look at the individual rancher. What's going on at Tony Perez's place? That's what's important."

Environmentalists, All

YOUNG ALBUQUERQUE ENVIRONMENTALIST Bill Bishop laughed when he remembered the exchange. "I was talking to a sheepman one day," he said, "and I suggested to him that the industry could solve its own problems if it would just develop a hybrid sheep that would eat coyotes. He thought that was a great idea, and he said, 'After they've eaten the coyotes, perhaps we could train them to eat environmentalists!'

"We were just having fun," Bill went on. "But there was something symbolic about it too. Because we seem to be polarized: the environmentalists and the coyotes in one camp, the sheepmen with their 'hoofed locusts' in the other. Neither side trusts the other. In fact there's hardly any communication between the two."

I mentioned to Bill some of the mail I had received: the letters from ranchers, some of whom damned all environmentalists as ivory tower biologists or "ignorant city folk from the East"; and the mail from "nature lovers" who view all sheepmen as congenital liars and who swear that if the coyote ever kills prey larger than rodents and rabbits, he takes only the halt, the sick, and the lame. I recalled one letter from a lady in Mesa, who wrote about the coyote's exemplary family life, and concluded that "they are very sweet animals."

Bill snorted. "I get just as frustrated," he said, "when I run into somebody who has what I call the 'baby puppy syndrome,' as I do when I talk to hard-core coyote-haters. The coyote is no 'baby puppy.' He's tough, strong, amazingly strong. He's a highly efficient predator. There's nothing 'sweet' about him—except perhaps to another coyote!

"But you get a lady like that together with a sheepman who knows the coyotes are eating him up and all they'll do is shout and never listen. And that's how we get into an escalating war of pictures and propaganda. 'All coyotes kill sheep.' 'All sheepmen are liars.' I show a photograph of a torn-up lamb. You show a photograph of a heap of poisoned coyotes. I show a picture of a cow who was attacked while giving birth, with her rear end all chewed up.

You show a picture of carbonized coyote pups dug out of their den after being burned alive. And where does that get us? Nowhere."

He'd been to too many meetings, Bill said, that deteriorated into shouting matches at which insults were traded rather than information. "And that's too bad," he said, "because we've got biological facts on our side which we've got to get the livestock people to see. Depredation is a site-specific problem. That's the biological reality, but the political reality is something different. What I find most distressing is that it's been impossible to convince the livestock industry in general that they need to look at depredation on a case-specific basis. They persist in favoring population reduction as the only thing that's going to get them out of trouble.

"And the industry *is* in trouble, particularly the sheep industry. The wool industry's in trouble because they don't have a marketable product. The demand is down on both wool and mutton. That's what's killing the industry, not the coyotes. But *they* say it's the lack of 1080 that's doing it. It's amazing that they aren't attacking the central problem, which is developing a market. Very rarely do you see TV commercials for wool—you see commercials for cotton all the time."

But surely, I said, there must be livestockmen around with some flexibility in their thinking, with whom one can sit down and have a rational discussion.

"You bet there are," he said. "Take Charlie Lee; he's president of the cattlegrowers in New Mexico. He's ornery, suspicious, but a man of great integrity. He's the industry's strongest advocate in the state. He cares strongly for the industry and the way of life it represents.

"He owns a beautifully-managed ranch near Alamogordo, and he deals in a concrete kind of conservation. He has a real, honest-to-God love for the land. He protects his land, he goes through hell to stop erosion if it shows up. If everybody in the industry were like Charlie we'd have no problem."

I arranged to meet Lee at a coffee shop in Socorro—he was driving north to attend legislative hearings in Santa Fe, I was driving south to keep some appointments in Las Cruces. Remembering Bishop's description of Charlie Lee as "suspicious and ornery," I was somewhat on guard when we met. But I found this tall, slender, clean-cut man courteous, obliging, and calmly thoughtful. For that matter, courtesy was a quality I encountered in all the ranchers with whom I spoke. Like Bill Bishop, Lee rued the lack of communication between the livestock industry and its critics. "Our industry," he said, "is just beginning to feel the public pressures that other industries faced long ago. We've been very independent and isolated. We never had to deal with public opinion in the past, and don't know how to do it now. When people criticize us we react with distaste and frustration. We tend to stay away from people who are giving us problems, when what we should do is meet them and tell them our side."

As for coyote depredations, they varied a good deal from spot to spot, Lee said. "I live in an area where historically there were many coyotes, but there were also many small animals for the coyote to feed on. But especially in the last five years there's been a definite increase in the coyote population, and an increase in calves bobtailed or killed. Now you see coyotes, seven or eight in a bunch, in broad daylight. You hear them almost every morning."

Lee did not like poisons much, but he felt they should be available for use when the coyote threatened to get out of hand. "Now we have no tools. There's a lot of trapping in our area; it brings people extra income and it's of some help to ranch operations. But it's not a very effective tool because if a coyote's escaped once he's smartened up. And the trap's a tough way to go. The coyote is just a dog. And I feel the same way when a coyote's in a trap as I would if the family dog were caught."

This was a very different attitude from that of the sheepman who told Dr. William Stephens, "When I get that coyote in a steel trap I like to leave him in it a while. I like to give him three or four days to repent his sins before I shoot him."

But "when it comes to the coyote," Lee told me, "cattle people are not nearly so rabid as sheepmen. Most of us, basically, are really environmentalists."

In so saying, Lee echoed a statement that was made to me by practically every rancher I met. "Some wildlife writers," John Olson said to me in Phoenix, "put us down as a bunch of wild animal killers. And yet a high percentage of cattlemen are preservationists. We're fighting the people in the industry who want to eradicate the last coyote. Most thinking cattlemen want only to control them."

As an *industry*, Olson said, "the cattle business has not been harmed by the ban on toxicants. But if you're only running 200 head of cattle you can't afford to lose *any* calves. A bigger operation can. This is a marginal business. The only real asset is the land, which is going up in value all the time, and you can't realize your profits on that unless you sell out to the developers."

Coyotes, Arizona Cattle Growers president Joe Lane said to me when he stopped by my motel in Tucson, had been only a "minor

problem for years. *Some* years they're bad. There's a relationship
between depredation and the number of rabbits. Right now we
have a huge population of both coyotes and rabbits."

I asked Lane what might be considered a "bad" year. "A loss to
predators of ten percent would be serious," he replied. "On our
place we've been losing from one-half to one to two percent of the
calf crop to coyotes. I have a neighbor who has twice my problem.
I'm not sure why—perhaps because his land is rougher than mine.
The number of bobtailed calves is indicative. If you find two or
three bobtailed you know that you've lost a few little ones. When
you have one cow, one calf, and one coyote, the coyote never gets
the calf. But you don't often have that situation.

"In our area predation is the biggest loss factor, followed by
lightning. But it's best to accept some loss to coyotes. Total control
is wrong: rabbits can do more damage than the coyotes. We need
periodic control. I say go back to 1080, let Fish and Wildlife handle
it, but don't put it out every year. Put it out every two or three
years, or whenever coyotes get too many. But there's many things
I worry about a lot more than the coyote!"

Indeed other matters were on the minds of the cattlemen at a
Tucson meeting a few days later of the Livestock Protective Association of Pima, Pinal, and Santa Cruz counties. The total silence on
the coyote was, I thought, significant. The problems that were discussed were problems that happen to be of common concern to the
cattle and sheep industries: the water-rights registration law,
grazing fees, how to elect more stockmen to the legislature. And it
has been suggested that the need to present a common front on
such matters has led the cowman to back his ancient enemy, the
sheepman, in his demands for more intensive predator control,
even though depredations may be of only minor concern to the
cowman. Thus in 1975, Arizona cattlemen had joined their wool-
grower brethren in supporting a measure that, had it passed,
would have allowed the issue of permits for shooting "predatory
animals" from the air.

But most sheepmen I met were as anxious as the cattlemen to
portray themselves as environmentalists. Two past presidents of
the New Mexico Woolgrowers Association, Phelps White and
Robert Naylor, came to the point the minute we sat down to talk.
"The important thing for you to know," White said, "is that
ranchers are conservationists *first*. As a group they would *oppose*
the elimination of the coyote."

"The sheepman has always had a problem with the coyote,"
Naylor said. "And with land values way up we don't have to raise
sheep. But we feel that what we're doing is important to the

nation. We're producing both meat and a natural fiber that doesn't cost oil to produce. But mainly we stay in because we love the business. It's been our livelihood."

White picked up the ball. "Ranchers are caretakers. In developing water sources we benefit the land, and we benefit the wildlife. And to some extent, whether we want to or not, we benefit the coyote, too. Coyotes kill for many reasons. They kill to survive, they kill to play, they kill to train their young. And it hurts us. But we're not out to control the coyote. We're out to protect our livestock."

The day after my talk in Roswell with Naylor and White, I drove on a raw, foggy morning a few miles out to Ellis Whitney's ranch. Whitney served four terms in the New Mexico legislature between 1945 and 1952 and has long been a powerful figure in stockmen's councils in the state. A tall, friendly, humorous man, with his very first words he echoed Phelps White: "Ninety-five to 98 percent of ranchers are conservationists," he said.

Whitney cited labor and predators as the local sheepmen's biggest problems. I asked him if sheepmen could accept the loss to predation of a certain percentage of their lamb crop as a cost of doing business. "Oh yes," he said, "I think so. The average man could live with a 5 to 6 percent loss. There are not too many of us who would eliminate the coyote—not that we could. The coyote can adapt anywhere and will survive. He's a predator, just like you and I. You hate them, yet it's a thrill to hear them howl. We wouldn't do without them."

And yet, in apparent contradiction of what he had just said, Whitney indicated that any coyote seen, or whose tracks were found, in the pastures was the object of an immediate hunt. "We run most of the coyotes on our ranch with pickups," he said. "We must have had three or four chases a week in 1973. Everything stops when you do this. You pile into the pickups, and go racing across the country after him at top speed, whatever the terrain. If you've got guts, not much sense, and don't give a damn about the pickup you're O.K. At least if you're lucky, you are. My partner broke his back doing this." But why, I wanted to know, go chasing off after any coyote whether or not it had killed any sheep? "When we find tracks of a coyote in the pasture," Whitney said, "we do not wait for kill signs. We eliminate him before he has a chance to kill. Any coyote confined in a pasture with sheep will be a killer in ten days to two weeks even if he has never seen a sheep before. Because it is sheep nature to run from anything that disturbs him, and it is coyote or dog nature to pursue anything that runs. And the end result is a dead sheep."

The net wire fences which divide the land—a checkerboard of

private, state, and Bureau of Land Management (BLM) lands—
into huge pastures came in in the twenties and thirties, Whitney
said. "They're a definite deterrent to coyotes. But now the BLM
wants us to take them down. At least they want us to replace 100
yards of net wire fence in every mile with smooth wire that will let
the antelope cross. The smooth wire will let the antelope through,
but it will also let the sheep through—all the ages will get mixed
up—and it will let the coyote through too. And soon there'll *be* no
antelope!"

Bill Bishop already had mentioned to me the net wire versus
smooth wire controversy. "The antelope herds are in real trouble
down there—several may be below critical level. With that net
wire fencing they have no way to migrate. But the woolgrowers
went out the window when the BLM proposed to roll back the
fences in spots to let the antelope pass. There has been a violent,
virulent reaction. They went at Secretary of Agriculture Earl Butz
about it last week. The truth is the sheepmen don't want more
antelope. They don't want hunters or sightseers or recreationists
on their BLM allotments. To hell with 'multiple use'! They're quite
honest about this. They don't say that antelope and sheep compete
for forage but that's a fact and they probably feel it! This is symptomatic of what you can expect from woolgrowers."

"The livestock interests in that area were always dominant in the
past," BLM chief biologist Ray Mapston said to me in Las Cruces,
"and they're not used to being crossed. But those fences they've
put up to control the coyote are also very detrimental to the antelope. We have removed sections of the net wire fence and the
sheepmen are very upset. They have this deeply ingrained belief
that the fences keep out the coyote, but it's not substantiated.

"There's a classic confrontation coming up," Mapston went on.
"As habitat managers, we're in a real dilemma trying to establish
multiple use. We manage 13½ million acres in New Mexico. This
is public land that belongs to all the people. It supports a public
resource, wildlife, that belongs to all the people. And yet there's a
big bureaucracy involved in killing a certain segment of this wildlife for the benefit of a few private permittees. Only 7 percent of
the livestock raised in the West is grazed on public lands. Perhaps
we ought to consider subsidizing ranchers for their predation
losses. But certainly the coyote belongs more on these lands than
exotic ungulates like sheep and cattle.

"BLM is taking a tough line on predator control," Mapston told
me. "We're asking Fish and Wildlife, where is the environmental
impact statement that was promised? And what about the National

Environmental Policy Act [NEPA] of 1969? Is it being adhered to? We've told Fish and Wildlife that we won't authorize predator control, except in emergency situations, on our lands until we're assured it's in compliance with the NEPA and that it will be managed in such a way that it won't be harmful to other multiple uses. I personally always hated 1080. It was banned, and properly so. It was widely misused, and was a disaster for many wildlife species.

"The biggest problem of the livestock industry," Mapston concluded, "is to modernize and to realize what's going on. For better or worse, traditional ranching is disappearing, and with it some of our traditional western heritage."

The same idea was put to me, in stronger language, by Defenders of Wildlife's "Steve" Johnson, in Tucson. "The livestock industry," Steve said, "is the last hold-out of rugged individualism. By the time they wake up to other values they'll be done for. For their own good it's time we got tough with the ranchers. The good ones won't go out of business. But the business methods of a lot of them are as outmoded as their attitudes on predators. And the livestock associations shield the bad guys. This leaves people no choice but to castigate the whole industry."

Not only professional conservationists like Steve Johnson are prepared to "get tough" with the ranchers, I found. A highplaced New Mexico Game and Fish official, who asked to remain anonymous, said, "Maybe sheepmen should accept losses on federal land as one of the costs of running sheep. If they're not willing to do so we should turn over the range to antelope, deer, and elk." Similarly, Robert Lerchen, retired veteran of thirty-two years with the Forest Service, said to me in Prescott: "High country is premium range—you raise heavier sheep, lamb more twins, have no insects or worms to worry about as you do in the low country. In return for the privilege of using this range, sheepmen must accept some predation loss as an operating cost. If they won't, we should keep their sheep out."

A similar process to one which had occurred earlier in Arizona is now taking place in New Mexico, Brant Calkin told me. "There's been a healthy decline in the number of sheep in this state," he said, "accompanied by a decline in the political clout of cattle- and sheepmen." And the consciousness of this decline, Bill Bishop said, was making the stockmen "paranoid about anything which they suspect as posing a possible threat to their influence or livelihood." One example of this paranoia was given to me by Norma Ames, assistant chief of Game Management for Game and Fish, in Santa Fe. "In 1973 Game and Fish proposed to extend its jurisdiction to *all* wildlife, not just game species," she told me. "But the legislature said no because of the fear of livestock people that we would protect the coyote."

This kind of reaction has put into question the sincerity of ranchers' claims that they are true conservationists. Conservation involves more than taking good care of the land—that is enlightened self-interest, no more nor less praiseworthy than the factory manager's scrupulous servicing of his equipment. Conservation is also a philosophy, an attitude which often demands the subordination of economic considerations to ethical ones. Viewed from this perspective many ranchers' claim to being environmentalists is more dubious.

"I'll tell you what conservation means to them," G. Corry McDonald, head of the Technology Utilization Program at Sandia

Laboratories and a long-time New Mexico resident and outdoorsman, said to me. "It means getting rid of all gophers and rabbits, shooting all the hawks you can, cleaning out the fence rows and hedgegrowth because you're so jealous of the use of water that you hate to see it used for anything that's not productive. And of course you've disrupted the coyote's food chain, and he gets desperate and will knock off anything that's not defended."

Two studies, conducted by members of the New Mexico Institute of Mining and Technology's faculty, provide fascinating insights into attitudes towards predators and towards man's relationship to other species. Dr. Christian J. Buys, associate professor of psychology, analyzed the responses of 384 ranchers to a questionnaire mailed in 1973 to 900 randomly selected New Mexico livestock growers possessing grazing permits on public land from the Bureau of Land Management. In his report, "Predator Control and Ranchers' Attitudes," published in *Environment and Behavior* (March 1975), Dr. Buys concluded that:

> The validity of ranchers' attitudes surrounding the predator control issue is less important than the prediction that ranchers can be expected to behave in a fashion largely consistent with their attitudes that predators inflict severe damage on livestock. They may be expected to resist numerous nontoxic methods of predator control, to be unreceptive to the agency in charge of a nontoxic program, and to lean toward use of poisons, even if illegal. In other words, the predator control problem may not be resolved for many years.

The other study was the aforementioned statewide series of forums conducted in 1974 and 1975 by Drs. Buys and Gary R. Olsen. Dr. Olsen, assistant professor of history at New Mexico Tech, recalled, "Most ranchers seemed convinced they were right on the edge of going bankrupt. The sheepmen were much the more virulent. Cattlemen were rather bitter but not nearly so hostile. They were more open. We even found there *are* some ranchers who think the coyote has a place in nature, and who are against poisoning. But most of the sheepmen wouldn't even listen."

In their report on the 1974 forums, *Perspectives on Land Use*, Drs. Olsen and Buys state:

> First, and most important in the context of contemporary debates about the place of man in the world and the cosmos, it must be stressed that most New Mexicans clearly believe

that human needs, real or imagined, rightly take precedence over those of other creatures. As one of the participants in the Santa Fe forum noted, "as long as man is in the center of the picture, current programs, however detrimental to the animal populations, make good sense." Only a few of the persons we met, invariably younger than the rest, seemed open to the idea that man is merely one of many creatures on earth with no special rights and privileges. Lynn White, Jr. and those who have recently been so vocal in their criticisms of the "man-centered" values of the Judeo-Christian tradition seem to be quite correct in seeing this as a most-prevalent value judgement in this part of the world.

Related to this rather narrow perspective is a pervasive ethnocentrism which again, at least for the majority of the participants, seems to rule out the possibility of examining and learning from other cultural experiences. For example, when the idea of the Buddhist reverence for nature and its possible relevance to any discussion of wildlife management was brought up in Roswell, we were simply told, "You won't find no sheepherder-Buddhists around here!" This simple rejection seemed, at least to those in the audience, to completely answer any such objections. The self-satisfied attitude which the statement represents seems to reflect the thinking of the vast majority of those we met and spoke with. Even the American Indians who participated (and from whom we hoped for a less man-centered view) ultimately had recourse to what might be called the Anglo orthodoxy as regards predators.

"Got Him!"  9

"I'M AFTER A pair of coyotes," Virgil Burns croaked in a voice that showed he was losing a battle with laryngitis. "A male and a female. I jumped them the other day, but they got away before I could get off a shot."

We were crawling in low gear, in his green Fish and Wildlife Service pickup, along the fence that separated two sheep pastures several miles northwest of Roswell. The morning was cool and fair; a light breeze blew in from the snow-topped Capitán Mountains forty miles to the west. All across the wide, gently rolling land, the only visible animal life was scattered groups of sheep and an occasional raven. Any coyote that chanced to be within range of a spent rifle bullet would be pretty easy to see, I thought. There were no trees, no bushes, nothing but close-cropped grass and mats of turpentine weed—the latter symptomatic of an overgrazed range. Only a few hollows, where there grew patches of tall sacatón, and an occasional eroded gully offered any cover.

"Have those coyotes killed any sheep?", I asked Virgil, a big, burly man in his early fifties and a government trapper for fifteen years. "I don't know of any kills," he replied. "But they will." "How can you be so sure?" I wondered. "Because they all do. If a coyote gets into a sheep pasture he may not kill for a while, but eventually he will. And they go for the fattest, heaviest lamb every time."

As we bumped along we kept our eyes fixed on the bottom of the net wire fence. Few coyotes jump fences—those that do are almost impossible to catch. Most of them, to pass from one pasture to another, find or dig a hole under the bottom wire. It was for such gaps that we kept our eyes peeled, while Virgil rasped a commentary in an increasingly hoarse voice.

"Northerners [apparently a variation of the more usual "easterners"] are against killing coyotes," he said, bringing the pickup to a stop. "And yet they like to eat lamb chops. It doesn't make sense." We got out, and checked a small depression beneath the fence. It was an old one, and the snare was in place—a loop of stainless steel wire fastened to the bottom strand of the fence. Unless it were very small, an animal trying to squeeze through the opening would drag on the loop; it immediately would tighten on its neck or waist or hindquarters, and the harder the animal pulled against it the more firmly it would grip him.

We got back into the pickup and resumed our patrol. "The bird-watchers are afraid we're going to clean out all the coyotes," Virgil scoffed as we made a ninety-degree turn around the corner of a pasture. "Hell, there's as many coyotes today as there was fifty years ago. We don't go killin' just for the fun of it. We try to keep 'em down so they don't bother the sheep so much. We don't go after some old coyote who's twenty or thirty miles away from the sheep and who's not bothering anything."

All the while I kept myself braced for the sight of a coyote—or a fox or a jackrabbit—garroted by a snare, but the snares, of which we must have checked two dozen in the course of the morning, were all undisturbed. Coming over a rise, Virgil pointed in the direction of an adjoining pasture in which a band of sheep were peacefully grazing. "That's the pasture, a big one, six sections, where they killed fifty-one lambs last year," he said. "It took us two months to catch the coyotes that was doing it. It was one pair with five pups. The lambs weighed around twenty pounds when the killing started; they were up to fifty or sixty when we stopped it. The coyotes would eat the heart and liver out and that's all they'd eat."

We were checking our last mile of fence when we scared up a jackrabbit, the first I had seen all morning. As it bounded away to our left Virgil said, "I never kill a jackrabbit any more, and I've talked several ranchers into quitting shooting them. Because that jackrabbit might save a lamb!" Twenty minutes later we were back at the truck stop where I'd left my car, and I thanked Virgil for the outing. "Sorry I couldn't show you more action," he said. I assured him that was all right, that I would probably see some the next day. And I was right.

Paul Wasson picked me up at my motel at 7 A.M. He had been a truck driver and a barber prior to joining the Fish and Wildlife Service some five years ago. As Paul drove, he treated me to a commentary on the coyote, the fundamentals of which I had begun to know by heart—although the details might vary from teller to teller. "Without exception, when you start lambing and you've got some coyotes around, you're going to lose some lambs," he said. "You'll never have all your lambs born alive. Of course the coyote is a scavenger. And so a coyote, who's perhaps never seen lambs before, will pick up a dead lamb. He finds it's sweeter meat than the jackrabbit and it's easier to procure. Pretty soon he starts in on the live ones. He begins with the small lambs, gradually works up. And it's a taught process. He mates with a coyote who hasn't learned to kill lambs. Monkey see, monkey do. Without exception a coyote will kill a lamb when he gets into a sheep pasture."

We turned off the Carrizozo highway onto a dirt road leading south. "In Texas one morning I picked up thirty-six dead baby goats. It was just wanton destruction. I can't see saving a bunch of damned coyotes that's hurting something. In sheep country there's no way you can have both coyotes and sheep."

We passed a ranch house on our left, and the road veered west, following an irrigation ditch. We stopped at a gate, I got out to open it, and noticed the sign nailed to the gatepost: DANGER! TRAPS — ¡PELIGRO! TRAMPAS. "The coyote gets bored so he kills," Paul resumed as we drove on. "They're like humans in that respect. The majority of crimes are committed by people who are bored."

I saw the raccoon as soon as it moved. It was fifty feet ahead of us to the left of the road, on the embankment of the irrigation ditch, and as it saw the pickup approach it tried furiously to pull free. But the trap, chained to a buried stake, gripped it firmly by a forepaw. Paul stopped the truck, we climbed out, and out of the back he pulled a large wooden tool box and a square of tarpaulin.

While I stood a few feet away, he walked up to the raccoon, which turned to face him, emitting a spitting growl. Paul tried to throw the cloth over it so he could reach the trap without being bitten, but the raccoon grabbed a corner of it and hung on. There was a brief tug of war, then Paul said, "That leg doesn't look too good to me. I think I'll have to finish it." He dropped the cloth, pulled out a .22 pistol, aimed carefully between the animal's eyes, which glared at him fixedly. He fired. The raccoon gave a jerk, then went into a long series of shuddering convulsions before it finally lay still. Paul freed it from the trap, tossed the body into some bushes, reset the trap, sifting earth over it and placing a dead twig over the sifted earth. This he sprayed with a few drops of yellow liquid from a plastic bottle. "Bobcat urine," he explained. "I collect it from a couple of cats I keep in a cage in the garage." He packed up his gear, we got back into the pickup, and drove on.

"The coyote can't help himself," he explained. "He can have all the natural prey in the world available, once he gets into a sheep pasture he's like a man who goes into town and sees a gorgeous woman. This man's going to look her up and down—and it doesn't matter if he's married to the most gorgeous girl in the world and she's sitting right next to him.

"Well, look what we've got here!" Paul exclaimed. A large ewe was standing to the side of the road, a front hoof locked in a trap. She stood docilely while I held her leg and Paul broke open the trap's jaws, then she trotted away. Paul re-set the trap, and we moved on.

We had been driving for some time now up a wide, level canyon with beautiful sheer cliffs on either side. "Look up there!" Paul said suddenly. Silhouetted against the skyline of the canyon's south rim stood three brownish animals with large, curving horns. "Barbary sheep!" I exclaimed. "Right," Paul concurred. "Game and Fish have introduced them around here and I guess they're doing pretty well." We passed two homesteaders' houses, their roofs partially caved in but their walls, built of stones neatly squared off with hand tools, intact. "They built to last in those days," Paul remarked.

The gray fox snarled and growled, as we came upon it, in a voice that to me sounded more like an angry cat's than like a canid's. It was a beautiful little thing, sharp-faced, soft-coated with handsome contrasts between the grayish hue of its back and the red tones of its ears, neck, and legs, a magnificent bushy black-tipped tail and delicate feet, one of which was badly mangled by the trap. Apparently mindless of its pain it tugged against the grip of the trap, backing as far away from Paul as the chain would allow it. After studying the fox for a few seconds Paul moved fast. He flicked his cloth over its head, grabbed it by the tail and lifted it off the ground, holding it stretched full-length so it could not turn and bite. While the fox's snarl grew in volume and pitch to a shrill scream of pain Paul took a razor-sharp knife, reached over and deftly severed the trapped forepaw above the wrist. Then he tossed the fox a few feet away into a bush.

The fox picked itself up and began to hobble slowly up the side of the canyon, which here was a rocky slope studded with juniper trees. Every few feet it would stop, hunker down, and look back towards us. Paul studied its tortured progress carefully. "I don't know about that one," he said. "If he doesn't keep going I think I'd better shoot it." I could not decide *what* I would do—put it out of its suffering, or let nature decide whether it would pull through or die slowly. The fox decided for us: it gradually worked its way up hill and finally disappeared over the rim.

Paul re-set the trap, squirted the bobcat urine on a tuft of grass just behind it, and we were on our way. "You know," he said, "there's no animal in the world I respect more than a goddam coyote. He's Number One in intelligence and native cunning. When you follow one coyote for months, and you use every trick you know, and he outwits you at very turn . . ."

Another fox! But this one seemed unhurt, although I was afraid it *would* hurt itself in its struggles before Paul could release it. However he succeeded, and I watched with joyful relief as it dashed away in graceful, airy bounds.

Soon after that we stopped for lunch. We ate our sandwiches on the bank of the ditch which carried irrigation water, diverted from the Rio Hondo, the whole length of the canyon. The sun shone benignly upon us; the sweet-scented air was still, the canyon quiet. Browning's lines popped into my mind:

> The year's at the spring;
> And day's at the morn . . .
> The lark's on the wing;
> The snail's on the thorn:
> God's in his heaven—
> All's right with the world!

Only I made up my own version: "The lark's on the wing/ The fox in his trap/ God isn't looking/ The world's full of crap!" I felt angry and disgusted and apprehensive. There were forty-three traps in this line, Paul had told me—he had set them three days ago. "I try to run my lines twice a week," he said. We still had a dozen or more to check. How many animals were lying out there right now, I wondered as we finished our lunch, patiently waiting for they knew not what? And what was the justification for it all? "I can't promise you that we'll have a coyote," Paul had said early in the day. "I don't know for sure that there *is* a coyote around. One of the ranch hands thought he'd seen some tracks. And there was a bobcat up the canyon last year. Never did get him—but he might still be around." Was *that* the "proof of need" that ADC liked to assure cynics they now demanded as a prerequisite for their services? But I kept these thoughts to myself. I was not out here to be critical of Paul, who was doing to the best of his ability the job he had been hired to do. But what a way to make a living!

Not long after lunch we found our next victim: a ringtailed cat, caught at a spot where the dirt road crossed a dry wash. The ringtail is no "cat" at all, but rather a small relation of the raccoon, with eyes rimmed by black and a brown- or black-and-white-banded, bushy tail longer than its body. Nocturnal and omnivorous, its diet

includes insects, plant material, rodents, and occasional birds. Ringtails make charming pets, and early settlers of the Southwest often had them about the home as expert mousers and as companions. The ringtail is comparatively rare: in all the weeks—months, collectively—I had camped out in the Southwest, I had never seen more than its tracks, until now. It was a pretty little thing, and meek. It made no protest or struggle as Paul whipped out his knife and with one quick slash cut off its injured paw; it just limped away into a thicket. "That one will be all right," Paul declared.

Next came a raccoon, released unharmed. Then another gray fox. This one must have stepped into the trap soon after Paul set it three days ago, for it was barely alive. It made no move as we walked towards it, it only watched us apathetically through glazed eyes. This one looked like the animals Dick Randall had described: its imprisoned leg was swollen to three times its normal size, and the swelling reached up to and over the shoulder. There was no doubt what had to be done about this one. Paul took out his pistol, aimed, and fired. The fox shuddered and kicked and shuddered and kicked. "Guess I didn't quite get you!", Paul muttered, aiming and firing a second time. The fox's pointed ears flicked forward, and it lay still.

"God, I hate to see those foxes caught," I said a little later. "They're such beautiful little things!" "Oh, I don't mind killing the *foxes* so much," Paul replied. "I've seen what they can do to a small lamb. But that's the trouble with do-gooders. They're mostly city folk, who have no idea what a wild animal does." I took the implied rebuke in silence. "Anyway," Paul added, "you don't have to *love* a job to do it."

Paul's job was done for the day, and we chatted about other things as we drove back to Roswell. I mentally totalled up our "score": a raccoon, a fox, a hognosed skunk, shot and killed; a fox, a ringtail, a raccoon, released with the loss of a foot; a fox, a raccoon, and a ewe released unharmed. Forty-three traps re-set and re-baited, now all ready to spring. And all for what? Because of an unsubstantiated report that a coyote's track had been seen in the area!

"That's one of our two worst problems—the non-targets," Vernon D. Cunningham, ADC's State Supervisor for New Mexico, said to me. "The other is the fact that a big part of our money comes from livestock association and county funds. This means that we at the Fish and Wildlife Service are not always free to do what we feel we should do, or not to do what we feel we shouldn't do. I'm always in a bind, trying to be fair to the rancher who's having losses and do my job, and not overkill.

"But those non-targets, that's something I hate! We get many more of them down there in the sheep country, where there aren't so many coyotes to begin with. Traps are the most non-selective tool we have. But we agree that *one* coyote in a sheep pasture is too many. How do we get him? I think there's too much environmental concern," he said, "to go back to full-scale use of toxicants. And that's good."

On the whole I found a certain ambivalence towards toxicants, including 1080, among the Fish and Wildlife Service people I met, whether they were employed in research or in ADC. I also found a fairly general distaste for traps. Dr. Knowlton, for instance, thought that compared with other methods poisons were not, *per se*, either good or bad. It all depended on how they were applied. The old method of application of 1080—fifty-pound chunks of meat imperfectly treated—had been terrible, he agreed. But small, individual baits might be an acceptable method, and one preferable to traps.

Bob Roughton tends to agree with this. He wrote to me:

I believe 1080 is humane and far preferable (in the form of widely scattered, coyote-proportioned drop baits perhaps) to traps and aircraft gunning. Death throes and convulsions are not necessarily accompanied by agonizing pain. You are perhaps more concerned about non-target and secondary hazards; believing these can be reduced to negligible proportions, I am more concerned with the biological, ecological, and economic absurdity of population suppression as a control method. As you know aerial gunning is useless for any other approach. Anyhow, for whatever reason, 1080 isn't and never will be a final solution. . . . It's just too bad it can't go down on actual disadvantages rather than hysteria. . . .

"Mistakes were certainly made with 1080," Terry Anderson said. "But I know that many state game and fish departments were not in favor of an outright ban on toxicants. The Executive Order was a political, not a biological, decision. The Cain Committee were experts, but they were not experts on the coyote," he added. Vern Cunningham's alter ego in Phoenix, Bill Rightmire, declared to me: "I am neither for nor against poisons. And it's the same with any other tool. You *can* minimize the trapping of non-target species. One of these days we'll have non-lethal tools that work—sterilizing agents, aversive agents, repellants. Even then there'll have to be some manipulation of the coyote population before these other things work."

The belief that a non-lethal method to thwart coyote predation on livestock might indeed be developed was expressed to me, to my surprise, by sheepman Phelps White. "I am convinced," White said, "that it's entirely feasible that an effective repellant will be developed in a very few years, and most of this killing will stop."

An enormous amount of research, much of it only begun since 1972, is under way on the coyote, sheep, and coyote-sheep relations. And an important, but far from exclusive, aim of this research is the devising or perfecting of non-lethal methods of coyote control. Donald Balser handed me in Denver a four-page summary, titled "Section of Predator Damage Research—FY 1976 Annual Work Plans." Under the headings "Damage Assessment," "Depredations Control," "Predation Ecology," "Behavioral Aspects of Predation," and "Socio-Economic Studies of Predators," it lists no less than sixty-four different study projects, to be carried out either by the Fish and Wildlife Service's own personnel or by different universities under contract. The target dates for the vast majority of these studies lie some distance in the future—as far away, indeed, as 1982. (For an authoritative, comprehensive, and recent report on coyote research, I refer the reader to Appendix A: Fish and Wildlife Service research biologist Dr. Samuel B. Linhart's paper, *Progress in Coyote Depredations Research*, presented at the Great Plains Wildlife Damage Control Workshop in Manhattan, Kansas, in December 1975.)

The sheep industry itself is the focus of much research. In the previously-quoted *Audubon* article, George Laycock described an interesting experiment:

> Just when it began to appear that the war between coyotes and sheepmen would continue forever, I met sheep expert Hudson Glimp. Dr. Glimp is a former research specialist at the University of Kentucky and the U.S. Department of Agriculture, and a former faculty member at Oklahoma State University. He lives in Wheatland, Wyoming, where he manages a budding program that has sheepmen around the world watching with interest. He wants to revolutionize the sheep industry and may well do so in a manner favorable to the beleaguered coyote. Glimp is general manager of the new $6 million Y. O. Ranch installation, which will have a "confinement-rearing system" for sheep. Plans call for the housing of 5,000 ewes; lambs will be raised indoors on a production-line basis and fed artificial milk made of soybeans. The world's most modern sheep-shearing plant will be located nearby; also a new slaughterhouse from which 800 lambs a day will be packaged for the retail market. A predator-proof fence will enclose the entire operation. "We will eliminate the two biggest threats to sheep, which are parasites and predators," Dr. Glimp declared. He believes that children will have to go to museums to see sheep unless American woolgrowers move in this direction.

The operation was a brilliant success, Dick Randall told me, but the Y. O. Ranch died, apparently because of insufficient financing. But it is expected to be resurrected.

Pending development of new methods, ADC does its best with the tools at hand, every one of which is objectionable in one way or another. The Cain Committee Report rated all available methods of control as to effectiveness, economy, safety (to man and live-stock), selectivity (takes only target species), specificity (takes only offending individuals), humaneness, and lack of environmental impact. No method was found to satisfy *all* these criteria. As Balser said to me, mixing his metaphors a bit, "We still have a way to go before we can accomplish our ideal—which is to stop the potential sheep-killing coyote before he has egg on his face."

One method which the Cain Report rated Fair to Very Good in all respects but one (economy) was aerial hunting. I arranged through Vern Cunningham to observe an aerial hunt on a big cattle ranch northwest of Magdelena, N.M. Early on a brisk February morning, I joined a small group of ranchers and ADC men stand-ing at the edge of the grass runway at the ranch near a parked Supercub. At 7:30 sharp, we heard the clattering of the approach-ing helicopter. It landed next to the airplane, and Ben Morris, the pilot, and Rex Norris, ADC sharpshooter, stepped out. Hardly taller than his 12-gauge shotgun and wearing prescription glasses, Rex did not fit my mental picture of a professional gunner, but, as I was to find out, he is a superb marksman.

The operation, directed by ADC district supervisor Wes Bon-sell, was organized like a military campaign. There were the ground forces—ADC district field assistants Arnold Bayne and John Foard, and ranch manager B. W. Cox, in the pickups. And there was the aviation—the B.A.F., as I came to call it, Bonsell's Air Force: the Supercub spotter plane, with Wes riding as spotter, and the gunship helicopter.

That first day I rode in Arnold's truck and followed the action at a distance. The strategy was as follows: The pickups would spread out, sometimes on the same ranch road, sometimes on parallel roads. John's and Arnold's vehicles had sirens mounted on them. We would drive along some distance, then stop. One of the men would turn on his siren for ten seconds or so, then abruptly shut it off and listen intently for an answering howl from the enemy. If an answer came, he would radio an estimate of the coyote's position to the B.A.F. circling overhead, the helicopter very low, the Super-cub 500 feet higher. It was Bonsell's task in the Supercub visually to spot the coyote which the ground forces had spotted audially, and then to direct Morris/Norris in to the kill.

Vern Cunningham had told me, "With this system we probably spot a third to a half of the coyotes in an area, and we're 90 percent sure of getting the coyote once it's spotted." By lunchtime I believed him. Arnold's hearing was little short of miraculous. He would sound his siren, turn it off, then stand a few feet away from the pickup, cupping a hand around an ear. I would hear nothing, except the drone of the plane and chugging of the helicopter a mile away, or the sighing of the wind in the junipers. But after a minute Arnold would step back to the truck, pick up the microphone, and say, "Hello, Bonsell." "Go ahead, Arnold." "Can you see me down here?" "Yep, I see you." "O.K. About one mile due west of me, at the foot of that flat-topped knoll. There's two of them there. They're still barking." "O.K., Arnold. We're on our way." And I would see the B.A.F. peel off in the direction indicated. Presently Bonsell's voice would sound again, a note of excitement in it. "I see him, Ben! Two hundred yards to your left! He's going south down that wash!" A few seconds would pass, then Ben Morris's voice, with the clack-clack-clack of the helicopter engine providing background accompaniment: "We're on top of him!" Clack-clack-clack-clack—a sharp CRACK!—then Ben's voice again: "Got him!"

Nine times that morning I heard those words, "Got him!" Nine gray-furred carcasses lay heaped in the yard outside the ranch workshop when we trooped in for lunch. We sat around the living room for a few minutes, trading details of the morning's hunt. "Do you know why that last coyote we followed down the hill moved so funny after he'd jumped into that draw?" Rex Norris asked Bonsell. "No, I figured perhaps it had knocked its wind out. It was sure flying when it landed!" "Yeah, well Ben checked it out. It had broken both its front legs!"

At lunch—a hefty, delicious ranch lunch of meat, potatoes, beans, vegetables, salad, rolls, pie, and lots of coffee—I said that it had been my understanding that cattlemen were generally more relaxed about the coyote than the sheepmen, but that since my arrival in Magdalena the day before I hadn't heard a good word spoken for the animal. "Oh, I don't think we're as extremist as the sheepmen," B. W. Cox replied. "We don't mind having a few around. But since they took away the poison from us the coyotes around here sure have multiplied. They raised hell with the calves last year."

Had he lost any yet this year, I asked? "No, not yet, but we haven't begun calving yet. We're just about to drive the cattle down into the pastures. That's why we want to thin out the coyotes —before the calves start coming. You don't have to worry," he added, "another month or two and there'll be as many coyotes on this place as there were this morning." Then he turned to Wes

Bonsell and said, "Sure appreciate what you guys are doing for us. It's sure going to help. Any chance you could be back here next month?"

"I'll have to check my schedule when I get back to Las Cruces," Bonsell replied. "I might be able to work you in." (He did work him in on March 5-6, and 17 coyotes were killed. In May, in the course of an aerial hunt near Silver City, Bob Gill—the Supercub pilot—and Wes crashed into a hillside. Bob was burned to death. Wes escaped with three booken vertebrae and facial lacerations.)

That afternoon was more of the same. Thirteen times I heard the triumphant words, "Got him!" Apparently these coyotes had not caught on to aerial hunting like the ones Dick Randall had known in Wyoming. There were all kinds of hiding places on this ranch— juniper trees, rocks, ledges, gullies—but the coyotes kept breaking from cover and running for their lives, which of course was fatal.

Back at my motel I tried to sort out my impressions. How would I rate this aerial hunting? It certainly must have been very expensive, and I as a taxpayer was footing part of the bill. Score a minus on that one. It certainly seemed effective—22 coyotes for the day. Score a plus. It was selective: no non-targets by this method. Score a big plus. I had not witnessed any of the kills, but to my knowledge all had been clean kills. Score a plus for humaneness. Specificity? Score a big minus.

By B. W. Cox's own admission, there had been no losses to coyotes yet this year. There were therefore no "offending animals" to be taken, to use the Cain Report's phrase. The twenty-two coyotes had been gratuitously eliminated, without trial or appeal, like so many rats. How many of them would have killed calves if they had lived? Who knows? Perhaps a dozen, perhaps one or two, perhaps none. Furthermore, since no one had the remotest idea how many coyotes there had been on the ranch to begin with, nobody had any idea what proportion of the population twenty-two coyotes represented. And nobody could foresee what the ecological consequences might be of removing that many from the particular environment.

One of the letters I had received had come from F. Robert Henderson, who is in charge of wildlife damage control for the state of Kansas. The operation I had witnessed this day was an exercise in population suppression, pure and simple—prophylactic control. Here, in part, is what Henderson had written to me about this type of predator control:

> . . . Man has wasted millions and millions of dollars to gain control of predator losses but he has thus far failed and most of the

thinking in this country is in error because it fails to recognize the real problem, and that lies with man's care of livestock and the attention that is required to prevent catastrophes from happening in respect to predatory animals zeroing in on easy to obtain food left unattended and abandoned by man. Each case of predatory animal reported loss to livestock should be examined separately and evaluated separately. In no case should any attempts be made to exterminate or even to control the population of predators in order to control the losses predators cause to man and his livestock. It should be noted, however, that in some cases the removal of some individual predators might be warranted, but unless livestock management is changed so as to take away the opportunity of the predator to kill again, killing again will surely resume and nothing short of changing the management practice will prevent it from happening again.

The next morning it was my turn to ride in the middle seat of the helicopter. I strapped myself in securely, made sure my camera was ready to shoot, and prepared for action. Ben lifted the craft a few dozen feet off the ground, then swooped forward and I found myself immediately gripped by the exhilaration of the flight. Skimming low over the ground, just clearing the tops of the juniper trees, slipping off to the right, sliding over to the left, swooping up, swooping down—there was a feeling of speed, of buoyancy, of grace in this flight which reminded me of the sensation of running a big rapid in my dory. We passed over a cliff-girt mesa, whose flat summit was covered with the foundation walls of ancient Indian dwellings. From the air I realized even better than I had from the ground what an incomparably beautiful land this was, with its ossature of red rock everywhere poking through the green epidermis of thin, grassy topsoil.

But I had little time to admire the scenery, for soon Ben Morris responded to a call from Arnold Bayne. I had no earphones on so did not know what Arnold was saying, nor could I hear clearly Ben's reply into the mike over the deafening clatter of the engine. So I just watched and kept my camera at the ready. We came off the top of a tableland and swooped down its western flank. Then I spotted it, shining silver in the glancing light of the early morning sun—the figure of a coyote dodging among the junipers. We closed in fast, Ben slowed his speed down to match the coyote's and kept the helicopter just to the animal's left so Rex could shoot out the right window. There was a loud report, the coyote turned a somersault and lay still. Ben set the chopper down a few feet away from the carcass, climbed down to the ground, walked over to the coyote, examined it, and walked back to the helicopter holding it up by one rear leg. This he showed me through the open door: it ended just below the hock, in a squared-off, thickened stump covered with black scar tissue which now was bleeding from the run. At some time in the past this coyote had lost its foot in a trap. "Poor two-time loser!," I thought, as Ben deposited the carcass in a basket fixed to the side of the helicopter.

It was all so ridiculously easy, so hopelessly one-sided. In this early morning light anything that moved over the ground—a rabbit, a small bird—was immediately visible. Even a half mile away a running coyote glowed like a tracer bullet. And once aware of this monster that was hurtling at them from the sky the coyotes never used their famous intelligence. There were a thousand places to hide, but no, they used their speed, their strength, their wind, their stamina in a mad desperate gallop that inevitably ended in a head-over-heels tumble as a load of buckshot slammed into their bodies. Two, three, four, five coyotes ended up in the basket. Only twice did Rex need to take more than one shot, when a coyote dodged just as he pressed the trigger.

The sixth and seventh were running together across a sagebrush flat. We were almost upon them when they dove down a gully that opened out onto a deep, straight-sided arroyo. Without breaking stride they made a 90-degree left turn and raced up its flat, sandy floor. We had them boxed in, it seemed; crack! went the shotgun and the rear coyote fell in a heap at the foot of the right wall. We left it lying there, apparently dead, to be picked up later, and continued in pursuit of the lead coyote. Just as Rex was about to fire the coyote dodged into a narrow gully that came in from the right. We lost sight of it for a few seconds. "Stay in there! Stay in there, and you might stand a chance!" I prayed silently. But no. Up the far end of the gully it emerged, out onto a flat. It was beginning to tire, and in a hundred yards we were on top of it. Rex fired, and the coyote fell, thrashing, onto its side.

I expected Ben would land immediately to finish off the wounded animal, but no. He had spotted yet another across the arroyo near the point where this one had turned into the gully. We flew back there, and hunted around for a couple of minutes, but it had disappeared. Back to the fallen coyote we flew, landing ten yards away from him. He was still lying on his side, twitching, jerking, his feet running in place. He lifted his head as Ben opened the door; his mouth worked like that of a dying fish. Whether he was screaming in pain or just gasping for air I could not tell over

the roar of the motor. For several long seconds he fixed his sharp amber gaze upon us, in a look that haunts me still. They he lay his head back down. His legs still were kicking, although more feebly, as Ben walked up to him. Ben raised his .22 pistol, pulled the trigger; the coyote's ears flicked forward, in a death reflex I remembered from Paul Wasson's fox. At last he lay still.

What had Dick Randall said? "Some people are soft-hearted and some aren't, I guess." I suppose I belong in the first category, for I found myself deeply moved by this animal's death. Perhaps because he had run so long, and so hard, and so far— and so futilely. Perhaps because it had taken him so long to die. Perhaps because of that lingering, indecipherable look. I decided I had seen all I needed to see. I suggested to Ben that since I was now out of film, and since somebody else might like to ride as passenger, he take me back to the trucks. He agreed, but first we must pick up this coyote's mate—I assume that was its mate—the one we had shot first when we flew up the arroyo.

For five minutes we searched up and down the miniature canyon. It seemed to us we recognized the exact spot where the coyote had fallen; but it was nowhere to be seen. Yet we could see no place where it could hide. And it had lain very still when last we saw it. Had it only been stunned, and revived after we left? Or had it found a hole, invisible from the air, in the wall of the arroyo into which to crawl, and lick its wounds or die? I will never know. Finally Ben gave up, leaving the mystery unsolved, and we flew back to where the pickups were waiting. I thanked Ben for the fine flight, and rejoined the ground troops.

An hour and three coyotes later, Bonsell decided to call it a day. The wind was coming up, blowing dust, which made the flying dangerous and the visibility poor. We returned to the ranch, ten dead coyotes in the back of a pickup. That brought the two-day score to thirty-two. Yesterday's catch lay skinned in a pile outside the workshop; the streamlined, hollow-bellied, sinewy shapes looked like so many crimson whippets. Gazing down at them with mixed horror, pity, and fascination, I remembered that powerful paragraph in Paul Horgan's history of the Rio Grande, *Great River*:

> On barbed wire fences, like symbols of the new order of affairs over the controlled range lands, dead, skinned coyotes were impaled in a frieze—twenty or thirty of them at a time. They were stretched in mid-air with a lean, racing look of

unearthly nimbleness, running nowhere; and their skulled teeth had the smile of their own ghosts, wits of the plains. In the dried varnish of their own amber serum they glistened under the sun. The day of unrestrained predators was over.

10

"Having Fun"

"WE'RE A LITTLE bit allergic to writers around here!" Robert Scott, past president of the Artesia Varmint Callers and Gun Club Association, Inc., smiled as he said this. But he was not entirely joking. "A couple of years ago," he explained, "this fat slob came out here from a big Eastern newspaper. He spent the whole weekend getting drunk at the bar, being obnoxious and asking stupid questions. Then he went back home and wrote this blood-curdling story about what a cruel sport this is, what a blood-thirsty bunch of murderers we varmint callers are!" Scotty smiled again and said, "You write a story like that and we'll find you and shoot *you!*"

Thus forewarned, I nevertheless accepted Scotty's invitation to accompany him and his partner, head high-school football coach Mike Phipps, the next day as they competed on the second day of the Ninth Annual New Mexico Varmint Calling Championship. We met at 6 A.M., and we lost no time in getting out of town: the best times for calling, Scotty said, are the early morning and early evening.

We crossed the Pecos River and continued east some twenty-five miles in the direction of the Caprock, into an area of rolling hills and brush-covered sand dunes known as the Loco Hills. Here we turned off the asphalt onto a dirt road—one of a maze of roads that led past pumping oil wells and an occasional windmill and stock tank. There was no sign of any other humans in this broad landscape. There were eighty-nine entrants in this calling contest, but they were spread out over an area of perhaps 2,500 square miles and so we had this corner, Scotty's favorite hunting grounds, to ourselves. Making a sweeping motion with his hand, Scotty said to me, "That's a lot of country you're looking at. And it's good coyote country. You never have to worry about us varmint hunters wiping out the coyote around here!"

The first coyote we saw saved us the trouble of calling it. We were bumping along a rough road when it materialized, seemingly out of nowhere, a hundred yards ahead of us and to our left, loping along a course that would have intersected ours. Scotty slammed on the brakes and Mike jumped out the door, leaned his .22-250 rifle across the hood of the Bronco, pulled the trigger and the coyote crumpled as if struck by lightning.

We walked over to where it had fallen; Scotty held open a green plastic garbage bag; Mike lifted the lifeless carcass with one hand and dropped it into the bag. I was struck by how small, almost frail, it looked, this first dead coyote I had ever seen from close up (the aerial hunt happened later). So this was it, the terror of the West, the monocephalic Cerberus of the rangelands! I suppressed an urge to laugh.

A few minutes later we made our first actual "stand." We took up positions about a hundred feet apart just below the crest of a hill, Scotty to my left, Mike to my right, each backed into a bush to blur our silhouettes. Below us stretched a wide, brushy swale, with another low hill rising beyond it. The sun was just up; the slight dawn breeze was losing its nip; the air was full of the song of meadow larks. We remained absolutely silent and practically motionless; to scan the field we moved our heads by imperceptible degrees from left to right and back. A coyote opened up with a short series of staccato barks from somewhere not far to our left; another answered from some distance away. Then they, too, relapsed into silence.

Suddenly the peace was shattered by a blood-chilling explosion of agonized screams. It was Mike, blowing into his call, cleaving the general ear with horrid shrieks, whaaaaah, whaaaaaah, WHAAAAAAAH, WHAAAAAAAAAH, crescendo, as the fraudulent rabbit felt the hawk's talons sink deeper into its flesh, then diminuendo, the cries becoming weaker, shorter, tremulous, plaintive, and ending with a whimper. Then total silence.

A minute passed. Tension built up. I gripped the camera in my lap, ready to jerk it to my eye. Nothing moved. I began to relax. CRACK! Scotty had fired—at what? Then I saw it—he had *missed*! The coyote was loping from left to right across the hollow below us, perhaps seventy-five yards away. This one obviously had never been shot at before, for it was moving at far less than full speed. CRACK! Scotty's rifle spoke again; a puff of dust kicked up just above the coyote's head; this seemed to wake him up, for as Mark Twain had described it, he "developed a livelier interest in his journey, and instantly electrified his heels." CRACK! Another shot, another miss! I couldn't believe it! (Scotty had dropped the rifle earlier; the scope was out of line.) Another twenty yards and

this coyote would reach cover. *Run*, I willed him, *RUN*! Then
CRACK! Scotty had made the necessary compensation; the coyote
tumbled over, gave three or four kicks, and lay still.

I expected Mike to congratulate him, but Mike was furious.
"The cat!", he cried. "Why didn't you shoot the cat?" "What cat?"
"Why, the bobcat! He was right in that line of bushes. I had my
scope right on him. I could have got him for sure! God, he was
pretty! If I'd known you were firing at that coyote I *would* have got
him." A bobcat earned twenty points in the championship, a
coyote ten, a fox five; hence Mike's annoyance.

The dead coyote was packed into another garbage bag and we
moved on, making at least another dozen stands. Scotty and Mike
alternated calling, and I studied their techniques so that I might
emulate them some day, camera in hand. Each had his own style.
And each had enough different calls tied to a string around his neck
to equip the wind section of a symphony orchestra. But it was all to
no avail. We saw no more coyotes, or bobcats or foxes either. I for
one felt no disappointment; I only hoped that my friends did not
feel I had jinxed them. If so they did not show it.

Scotty seemed philosophical about the poor hunting. He talked
to me about the techniques and philosophy of varmint calling. "We
hand-load our ammunition," he explained at one point. "It's
cheaper than buying it, although that doesn't save us any money—
we just get to shoot more bullets. But it also gives us much better
accuracy. And we use a bullet with a very light, thin jacket that
explodes on contact. If it hits even a twig it bursts apart. With
these bullets there's little danger of ricochets. And they're more
humane. When the bullet hits an animal—even a small, thin-
skinned animal—it explodes all through it. It's an almost certain
sudden death. You hardly ever have any cripples."

At another point, Scotty reasoned: "We think that our hunting
helps the ranchers out. We're helping to keep down the number of
coyotes. But what we're after is control, *not* extermination. If we
find a species getting scarce, we ease off on it. Bobcats are not too
numerous, but they're also very hard to get. But the fox—there
were a couple of years there where we voluntarily decided to drop
the fox from the competition—we gave no points for it. That was
because we felt that its numbers were too low, that we ought to
give it a break."

The actual kill, Scotty said, was the least important part of the
game. "The fun is in the calling, seeing what kind of animal comes
in, how close you can bring them in. I've had them land right on
top of me from behind. The actual shooting is anticlimactic."

"But then why *shoot*?" I asked. "Why the *kill*? What's the point? Especially if even *you* feel it's an anticlimax." This Scotty could not answer; at least he had no answer that made sense to me.

I turned to other hunters for the answer—Dan Caughey, a young sheriff's deputy in Honda, California, for one. We met in Honda, and he took me up into the beautiful green hills. We spotted a coyote lying on its stomach in a meadow at the edge of a wood. Dan let the car roll forward out of sight; we quietly got out and crawled back to where we could look down on the coyote, a couple of hundred feet below us. Dan blew his varmint call; the coyote leapt to its feet, looked up in our direction, crossed the meadow towards us at a run, then at the foot of the hill it disappeared from view. We never saw it again. "It most likely circled around us and got our scent," Dan conjectured. Dan had brought no rifle with him. "I only call to shoot during fawning season," he explained. "I think by taking a few coyotes then I may help out the deer."

I found, as I talked with Dan, that in matters concerning wildlife and conservation in general, we had many more areas of agreement than disagreement, even though he was an avid trophy hunter. He deplored the "slob" hunters even more bitterly than I have, and let pass unchallenged my statement that if we could not weed out the slobs we might have to ban hunting altogether. But still there was this matter of the trophy, of the kill. Why go that final step? His answer left me as unenlightened as Scotty's.

It was the same with Dan Cady, former president of the Utah Varmint Callers Association, who, in a long letter full of interesting anecdotes on calling, invited me to a meeting of his group in Logan. "The challenge is in outwitting the coyote, in getting him to come in," Dan said to me at the meeting. "I appreciate that," I said, "but haven't you already proved your point when you've called him in? So why go on to the kill, why shoot an animal you can't eat, unless perhaps it's been eating your sheep?" "I guess it's an instinct," Dan said. "Some people seem to have it, some don't."

I had a number of letters concerning varmint calling and trophy hunting. Some were rather strongly worded. Karen Kuykendall, for instance, a professional artist in Casa Grande, wrote in words reminiscent of Oscar Wilde's description of "the English country gentleman galloping after a fox—the unspeakable in full pursuit of the uneatable":

I think it's tragic that fine four-footed creatures must sacrifice their lives so that two-footed critters can display their heads and hides on the walls of their dens. Some men cannot feel "manly" without a gun in hand and an animal lying prone at the end of a smoking barrel. The whole machismo philosophy is silly and childish anyway, but when animals have to be slain in order for some Great White Hunter to be able to "prove" what a big, brave, and virile man he is, then this becomes just as criminal as the taking of a human life. The animals have every bit as much right to life as people do, and, frankly, I'd rather see a lot more animals in this world and a lot less people. The animals are far less trouble—and they don't start wars.

"Sport" and "trophy" hunters are among the lowest forms of humanity in my opinion. Whenever I see a photo of some grinning idiot squatting above a prone cougar or mountain sheep, I just cry inside and wish their positions were reversed. Each time an animal is slain, something very fine and noble departs from this world. I may never see a bobcat or a wolf in the wild, but just knowing that they are THERE *gives me a beautiful feeling. The animals learned to live in harmony with their natural environment long ago; it's high time people also were taking some lessons!*

A sometime (or sometimes) varmint caller, Michael Pottorff, wrote from Costa Mesa, California:

I am no longer as enthusiastic as I once was. I still have my calls and use them once a year, but 1. as I see coyotes within range of my bow 3-4 times a year while hunting deer I do not feel compelled to use my call, 2. putting something on the wall is not nearly as imperative to me now as putting something in the freezer, 3. I'm not sure that I'd disagree that it's cheating, 4. most importantly, I resent being associated with the mentality of those who, unlike me, are very successful at calling. These individuals seem to embody all those traits that I have associated with used car salesmen and with CIA operatives. My biggest turn off came when I read the result of the yearly kill of the La Mesa calling club. The winner and the next 5 finishers had scores that looked like this: 29 coyotes 15 bobcats 2 badgers etc. It was sick. . . .

The most articulate *defense* of varmint calling came from John Heuston, an outdoor writer in Little Rock, Arkansas:

As I've grown older [39], my philosophy of hunting has modified somewhat. I now no longer shoot any animal that I do not intend to consume for food. I derive just as much pleasure from shooting a photograph of a called-up coyote as I do killing it. The more I predator hunt, the more my respect, love and admiration for these animals grows. And the more I become convinced that they are sporting animals of the highest quality and deserve the status of game animals with seasons and bag limits.

Please don't misunderstand me. I have no quarrel with those who do want to call up "varmints" and shoot them if they have no personal qualms about doing so. I have nothing but contempt for anti-hunters and their over-urbanized attitudes and am firmly convinced that if these neurotics have their way wildlife will disappear from the North American continent in a few decades. This is not 1790, or even 1935, and without hunters to finance the maintenance of wildlife habitat and conservation-education programs there will be no economic incentives to keep land open for predators or prey. Then wildlife habitat will fall prey to the greatest predator of them all—man, his machines, and his insatiable desire to pave and subdivide every last inch of this continent. The bulldozer doesn't give natural selection a chance to work its miracles. Darwin never had to cope with the U.S. Army Corps of Engineers or the Farm Bureau Federation.

In my experience, predator hunting holds no threat to the survival of bobcats or coyotes. First of all there just aren't that many predator hunters. If the U.S. Government, with its massive resources, can't wipe out the coyote (which has adapted to all counties in Arkansas in the last two decades) the sport hunter certainly isn't going to be a threat.

In fact, have you ever considered this intriguing idea—the more predator hunters there are, the more sportsmen you have around that are concerned about the fate of their favorite quarry? They could, if organized, be a potent force in eliminating wasteful and non-selective government trapping programs in favor of sport hunting for these same animals.

Personally, however, I would like to see electronic predator calling machines—records or tapes—outlawed and calling restricted to mouth-operated calls. This puts an element of skill into calling that is otherwise lacking. Also, predators should be protected during breeding season. . . .

And why does man want to hunt and kill the animals that he admires more than any living things? I dunno. Maybe being locked

up on the 29th Floor of a high-rise business office in a plastic world
for five days a week has something to do with it. Maybe I want to
reassure my inner self that I really am a predatory meat eater and
that my meals haven't always come wrapped in white paper pack-
ages killed by someone else. For some reason, that seems very
important to me.

For those of you who might wonder, I will state that I have shot
and killed two animals in my life. The first time I was twenty, a
freshly commisioned officer in a French cavalry (tank) regiment. I
went out one evening with a young fellow-officer onto the firing
range. We each bagged a hare, and took our game over to a neighboring farm where a few days later the farmer's wife presented us
with the most delicious *civet de lièvre* (stewed hare) I have ever
tasted in my life. I have never had any qualms about that hunt.

The second time was a little more than a year later, when a
fellow graduate student and I spent a week end at his family's cabin
in Rhode Island. On a misty fall morning we sallied forth to hunt
some woodchucks. I had always loved the country, and was
amiably disposed towards animals in general, but I had yet to
develop a deep ecological consciousness or conscience. And so
when we spotted a woodchuck sitting atop its burrow near an
abandoned barn, and Henry invited me to take the first shot, I did.
As I remember it it was a shot of fifty yards; I hit the woodchuck in
the head. We walked over to it, I picked up the little body, once so
perfect in its way, and suddenly I felt ashamed. Who had given me
leave to take this life that I had had no hand in creating? What had
I proved—that I was a good shot? I knew that already, and if Henry
had required proof I could have shot at beer cans. That I was
manly? I felt the less of a man for having killed so callously. I did no
more shooting that morning, and I swore to myself that never again
would I take the life of an animal that I could not eat.

Thus my preoccupation with the "kill" in varmint calling, my
persistent efforts, futile as it turned out, to understand the
psychological motivation behind it.

I talked about this subject with Dr. James Findley, University of
New Mexico mammalogist. "West Mesa," Dr. Findley recalled,
referring to a plateau that overlooks the Rio Grande valley just
west of Albuquerque, "used to be a marvelous place to see all kinds
of predators. Now it's crisscrossed with roads, and you can't find a
thing. The varmint callers have done it. . . . Now all you find is
piles of expended shells."

The cult of guns and shooting, so prevalent in, but by no means exclusive to, the West, was but one facet, Dr. Findley thought, of the childish but dangerous ethic of machismo. "Machismo is a symptom of arrested maturation," he declared. "But it really makes you wonder, how can all these guys be so unsure of their masculinity?"

Officials of the various wildlife agencies with whom I spoke in Arizona and New Mexico did not seem to engage in this line of speculation. Those who mentioned varmint hunting at all, saw it as a commendable sport. Arizona Game and Fish Director Jantzen stated it flatly: "We are in favor of the sport hunting of coyotes. It's a speciality-type hunting, with a year-round season, and we definitely support it." Ray Mapston thought the same: "We should think of the coyote as a sporting animal," he said. "I would much rather see sportsmen hunting him as game than see him destroyed by paid government trappers."

I told Ray I might agree with him on purely tactical grounds, and I quoted to him Heuston's argument that "the more predator hunters there are, the more sportsmen you have around that are concerned about the fate of their favorite quarry." But ethically and philosophically I could *not* agree. "Killing an animal that you do not need for food seems to me a perversion of the hunting instinct," I said. "If some animals must be killed for economic reasons, let it be done coldly, professionally, as an unpleasant but unavoidable necessity. But for God's sake let's not glamorize it," I pleaded. "You'd only be encouraging the idea that it's fun to kill. And that's already much too popular an idea."

Mapston looked at me quizzically: "I've never looked at it *that* way," he said.

There was no doubt whatever that the varmint hunters gathered in Artesia were having fun. The championship ended at 4:30, at which time all entrants had to present their catches for judging, and the parking lot behind the restaurant where the banquet would later be held was jammed with pickups. The contestants milled about, laughing, teasing, trading anecdotes, and admiring each other's haul. In the back of one pickup were piled nineteen coyote carcasses—this team won the first trophy. In another, eleven coyotes and two lovely little gray foxes. This team came in second.

The grand total was announced at the banquet: an all-time record of 104 coyotes, 8 bobcats, and 6 foxes. Loud cheers. The Artesia Varmint Hunters president declared: "The whole idea of varmint calling is getting outside and having fun." More cheers. The winning teams were called forward one by one to receive their trophies. Cheers and loud clapping.

The banquet hall was reminiscent of a high school locker room after a win. I had to keep reminding myself that most of the celebrants were not boys but adult men, and that the statistics being bandied about had to do not with touchdowns or baskets or home runs, but with lives—or what *had* been lives at the start of the day. Every one of those creatures frozen in rigor mortis in the back of the pickups had been alive this morning, a sentient being with its own fears, its own wants, its own preoccupations, its own pleasures. I wondered if anybody in the banquet hall that night ever thought of those animals in these terms. The attitude seemed rather to be that life, because self-renewing, was casually expendable.

And yet, in talking with them individually, I found much to like and to respect in these men. There were some sheepmen among them who were coyote-haters, but there were at least as many men who admired and even liked the coyote. And I found some real ambivalence in my reaction: the very idea of coyote hunting is repugnant to me; deer hunting is not. But I would far sooner trust a Dan Caughey, a Dan Cady, or a Robert Scott with a rifle than I would many deer hunters. The guest speaker this evening was Game and Fish director Bill Huey. At one point in his talk he declared: "There is a significant number of deer killed each year and left to rot in the fields. We blame it on the 'slob' hunters and do nothing about it. This plays right in the hands of those groups, particularly back East, who would like to see hunting banned altogether. We must begin to identify the slob hunter and do some talking to him!" This remark was greeted with loud applause.

I have spent a good part of this chapter discussing varmint callers, but I suspect that the coyote's welfare is potentially far less threatened by the varmint callers than it is by run-of-the mill deer and elk and antelope hunters who press for coyote control. To the varmint caller the coyote is prey; to the deer hunter the coyote is a competitor. Few hunters, human or otherwise, hate their prey. But a competitor is another matter.

I remember a few years ago asking a bartender in Flagstaff, whom I knew to be an avid hunter, "When you're out hunting deer or elk, and you run across a coyote or a bobcat, what do you do?"

"I shoot him, of course."

"Why?"

"I figure he's no damned good. If you left it to the coyote, there'd be no deer around."

"Do a lot of your hunting buddies feel the same way?"

"Oh, I think just about all of us do."

In Santa Fe, Bill Huey said to me, "The majority of sportsmen are convinced the coyote does significant damage to wildlife." Many of the letters I received from hunters all over the West corroborated this.

There was near unanimity among Game and Fish officials in both New Mexico and Arizona on the impact of coyotes on game herds. Bill Huey put it this way: "Sometimes the coyote does significant damage, sometimes he doesn't. It depends on the status of individual species. If it is in a critical status, then the coyote can exert pretty serious control."

Bob Jantzen told me in Phoenix, "Low deer or antelope numbers are not necessarily tied in to 'too many coyotes.' But there are times when the coyote can hurt other wildlife resources." Again, Ronald Smith, Game Research Supervisor for Arizona Game and Fish, explained that "when a herd gets down to a threshold level the influence of predators may then control the population below the capacity of the range."

Studies are under way in both states (on the deer at Fort Baird in New Mexico and the 3-Bar study area near Roosevelt Lake in Arizona, and on the antelope on Anderson Mesa near Flagstaff) to determine the relative importance of predation and other mortality factors on herds of game animals. But it seems to have been factors other than predation that caused the herds to shrink in the first place, and that contribute, along with predation, to hinder their increase today.

According to Bill Huey, fire suppression "has been the most critical factor limiting browse for deer, and has also affected the elk herd. The emphasis on grass production has been good for live-stock, but bad for deer." Another official phrased it in even broader terms: "There's an enormous amount of human distur-bance of wildlife, which is especially critical during breeding season. Wildlife is being nibbled away from a multiplicity of causes."

In the 1800s, according to research biologist Don Neff, "all of Northern Arizona from Leupp to Needles was marvelous antelope range. Then it was sheeped to hell. Added to that, in the 1950s hunting and poaching had a horrendous impact." "Why were there so many antelope 100 years ago if the coyote is such a villain?", asked A. Steve Gallizioli, Chief of Research for Game and Fish in Phoenix.

Human harassment of wildlife can be quite inadvertant, as Neff made clear in discussing the Anderson Mesa antelope herd. "If a cow gets run off by a pickup or some other disturbance, the fawn is then fair game for the coyote. And this happens all the time—spring fishermen coming in, cattlemen checking their fences, all kinds of things. These antelope are awfully spooky. They'll run when you're a half mile away."

The pattern seems clear. Human disturbance or destruction of the habitat, rather than coyote predation, has been the direct or indirect cause of the decline here and there of certain game animal populations. But in the eyes of many hunters, the coyote (or cougar) is the culprit. And he *may* be at least in part responsible, in certain instances, for the failure of game species to rebound.

And so the pressure is on the game and fish departments to control the coyote so that there will be more deer to shoot. Bill Huey emphasized that "we try to base our decisions on biological considerations, not on political ones. We have the welfare of *all* wildlife at heart, and we consider the coyote to be a highly valuable native species, an integral component of the ecosystem." But another official admitted, "deer are our bread and butter."

Bob Jantzen conceded, "We manage the coyote in favor of the antelope. We do it primarily for the benefit of the hunters, although tourists profit, too. The antelope is a highly visible animal, the coyote is not."

Bill Rightmire, it seemed to me, went to the core of the matter. "It's up to Game and Fish," he declared, "whether it wants to beef up the antelope herd. And they have a philosophical, as well as practical, choice to make: which predator will they favor, man or the coyote?"

One who would not agonize much over the choice, I gathered, was Elliott Barker, who headed the New Mexico Department of Game and Fish for twenty-two years before his retirement in 1953. Still vigorous at 89, Barker received me at his home in Santa Fe. "Control is the thing in all wildlife management," he declared. "Too many deer is just as harmful to deer as anything else could be. The public has got to come to the same realization where mountain lions and coyotes are concerned. We need control, control, control," he said, thumping his fist down on the arm of his chair.

"Sportsmen should be entitled to two-thirds of the deer harvest each year," he continued. "You have to spare some for the predators. If you maintain the hunter kill at 30,000 deer a year, you can't afford to give more than 15,000 to the lion. One lion kills an average of a deer a week. So to assure sportsmen of their two-thirds share you've got to control the lion population down to 300." (There may be more than 2,000 lions in the state.)

By what logic or ethic, I wondered as I listened on, could one decide that "sportsmen should be entitled to two-thirds of the deer"—if to ensure this allotment hundreds of lions must be killed? Why should the *lion* not get the two-thirds, or even the totality, of the deer harvest if he needed it? Few Americans today are economically dependent on game for survival, and if there are such unfortunates there are other ways to help them than to decimate competing predators. For nine hunters out of ten the pursuit of deer or elk is a recreation, a diversion, a way of "having fun." For the lion or coyote, prey is a necessity of life. Who, then, has the more serious claim: the mountain lion, wolf, coyote—or man?

Jack Berryman, then Chief of the Division of Wildlife Services, made a revealing statement in a speech in 1970: "We can indeed manage wildlife to minimize conflict with man—to assure its future."

I look forward, but with little hope, to the day when a man in Berryman's position will turn the statement around and say: "We can indeed manage ourselves to minimize conflict with wildlife—to assure our sanity."

One for All

There slowly grew up in me an unshakeable conviction that we have no right to inflict suffering and death on another living creature unless there is some unavoidable necessity for it.

—Dr. Albert Schweitzer.

The attitude back East is very different from here. Easterners react differently from us toward wildlife, toward life and death. There is a tremendous development back there of a tendency to want to isolate oneself from death. Out West we know it can't be done. We accept death as part of life. Something has to die so another creature can live. Hence our attitude toward hunting is different.

—William Huey.

AS I LISTENED to Bill Huey speak these words to his appreciative audience of varmint callers in Artesia, I wondered how Dr. Schweitzer would have reacted. Would he have wondered, as I did, if the contrasting of East with West was not a trifle simplistic and demagogic? Would he have questioned, as I did, the logic of equating acceptance of the inevitability of death with willingness to be its agent? I suspect that he would have.

I suspect that he might also have questioned a statement I often heard made in defense of game management, that cropping the deer herds was a kindness to the deer: "Starving to death or being eaten by a predator is so much more terrible than being shot." What would the *deer*, or elk, or mountain lion, or whatever, say if they could be asked and could answer the question, "wouldn't you rather be shot right now while in your prime than take the chance of an ultimate death by predation or starvation?" What would Bill Huey, or you or I, answer if offered such a choice: euthanasia now, or cancer later?

My quarrel is not with the hunters as such—certainly not with the majority of decent ones. Even less is it with the men and women of Game and Fish, whom I like and respect. Rather, it is with our common human arrogance in wanting to play God with the destinies of other creatures. What right have we, I wonder, to "manage the coyote in favor of the antelope"? When we manage a species—by harvesting so many elk, or by killing 81,000 coyotes as Animal Damage Control did nationwide in 1975—how accurately can we predict the results? How often, when we play God, do we really know what we are doing?

"There is a story," Loren Eiseley writes in *The Firmament of Time*, "about one of our great atomic physicists—a story for whose authenticity I cannot vouch, and therefore I will not mention his name. I hope, however, with all my heart that it is true. If it is not, then it ought to be, for it illustrates well what I mean by a growing self-awareness, a sense of responsibility about the universe.

"This man, one of the chief architects of the atomic bomb, so the story runs, was out wandering in the woods one day with a friend when he came upon a small tortoise. Overcome with pleasurable excitement, he took up the tortoise and started home, thinking to surprise his children with it. After a few steps he paused and surveyed the tortoise doubtfully.

"'What's the matter?' asked his friend.

"Without responding, the great scientist slowly retraced his steps as precisely as possible, and gently set the turtle down upon the exact spot from which he had taken him up.

"Then he turned solemnly to his friend. 'It just struck me,' he said, 'that perhaps, for one man, I have tampered enough with the universe.' He turned, and left the turtle to wander on its way."

Contrast that attitude, if you will, with the former Game and Fish director's fist-pounding apotheosis of "Control, control, control!"

To control or not to control. The coyote and the sheepman. The lamb killer and the government trapper and the varmint caller. These are not easy subjects to discuss impassively. And indeed emotion long has dominated the discussion on both sides. Both the animals, and their human enemies, have been described either as epitomies of cruelty and depravity, or as models of self-reliance and nobility. They have been viewed as heroes or villains, in a drama in which violence is the common mode of action.

And yet, although I, too, have stong feelings on the subject, I see few real villains in the story, unless they be the likes of those men and women who find sport in pursuing coyotes in jeeps or snowmobiles and running them over. The calf-killing cougar, the sheep-killing coyote, the antelope-killing golden eagle, the moose-killing wolf of the north are not villains: they react in response to individual needs and environmental imperatives over which they have little, if any, control.

The average Animal Damage Control district field assistant is no villain either—some may lack imagination, biological wisdom, or empathy for their victims, but they are generally convinced that they are performing a needed service and carry out their duties to the best of their abilities. Nor is the stockman who seeks the eradication from the range of such competitors as the coyote or mountain lion, a villain. He *has* lost animals to predators, and rightly or wrongly he concludes that only by eliminating the predators or drastically reducing their numbers can he remain in business.

But one does not require an extensive knowledge of human history to realize that it is not only villains or psychopaths who commit atrocities. It has always been easy to justify evil by invoking necessity, self-defense, or even nobler compulsions—and all the more so if the victims are animals who are supposedly bereft of anything approaching the human capacity to think and to feel, or who, alternatively, are credited with powers of premeditated malice and cruelty that are, in fact, peculiarly human.

Historically, our treatment of predatory animals has been a long tale of atrocities. Even today, many die cruel and often unmerited deaths, like that raccoon and that fox I saw in the traps, or that agonizing coyote who had run such a gallant race and whose yellow eyes still stab into my soul in the small hours of a restless night.

On the other hand, as George Merrill of New Mexico Game and Fish said to me, "Cruelty to individual animals—that's what you sometimes see in control programs. But that does not mean that a *species* is being hurt." As far as the coyote is concerned, I would say Merrill is right. There seems to be no doubt at all that as a species, the coyote is "alive and well." He is alive because he is smart, he is tough, he is adaptable, he is prolific, and because there are large areas where there is little or no coyote control. And these include the outlying sections of big cities such as Albuquerque and Tucson, where household pets sometimes supplement the coyote's ordinary diet. (My friend Sidney Franklin told me recently that early one morning, while driving along Skyline Road in Tucson, she had seen a coyote cross the road, holding a dead German Shepherd puppy in its mouth. Had that puppy been my Hamlet when he was young, I should have felt, at least for a while, as hating of the coyote as many sheepmen do.) And yet it is not the coyote who has invaded the cities, rather it is the cities that have sprawled out into coyote territory.

I stayed ten days last March at a motel on Oracle Road in Tucson, and nearly every night I heard the excited yipping of coyotes as they chased rabbits across the golf course below. "¡Cantad, amigos!" I cried more than once. One night of a full moon, feeling exuberant, I stepped out onto my balcony at about two o'clock in the morning and blew three or four loud squawks on my Weems varmint caller, then stepped back into the shadows to await results. No coyotes appeared, but there was a precipitous tumbling out of their rooms by other motel guests, shouting "What was *that!?*", "Are you all *right?*" "Is somebody hurt?" "Should I call the police?" I confess that I felt about ten years old!

By the end of my tour of investigation of the coyote, I became reassured as to the present, and probable future, status of the animal. But I remained convinced that the controversy over *Canis latrans* is still far from settled. Many environmentalists still deplore the continued killing of predators. Many stockmen deeply resent the ban on toxicants, especially on 1080. And ADC decision makers are in the uncomfortable position of being damned if they do and damned if they don't.

When I checked in at the ADC office in Albuquerque after observing the aerial hunt, Vern Cunningham asked me, "Do you have any suggestions?" I felt that his interest was genuine, that he was not merely trying to put me on the spot. But I could not help him. I had seen things that shocked me, yet I could offer no alternatives. I told him of my revulsion at the killing of thirty-two coyotes who, whatever their potential for mayhem might have been, were guiltless of any depredations at the time of their executions. "You know," he said, "a cattleman west of here called me just this morning, and he made much the same objections. He was really mad. I guess our helicopter strayed over his land by mistake. 'How do you know how many coyotes there are around before you start?' Of course we don't, and we *should*, but how do we get such figures?"

What then *is* the answer to coyote damage? Go back to the broadcast use of toxicants? God forbid, although I am not at all sure that one "tool," the M-44 cyanide gun, is not vastly preferable to the steel trap. But Donald Balser assured me, "The Federal government and the public won't put up with a return to the situation that existed prior to 1972—at least not until all the research is in!"

Ought we to listen to cattleman John Donaldson, Jr., who said to me in Tucson, "If we just let the coyotes alone they would soon control their own numbers"? And of course they would—as the U.S. Forest Service's Dale Jones, Region 3 wildlife director, put it, "If coyotes were left to their own devices, they wouldn't multiply to the point where you'd be kicking them down the street." But would a higher coyote population, controlled only by their own territoriality and the cyclical ups and downs of prey populations, result in greatly increased depredation on livestock? No one yet seems to know.

Are changes in management the answer? Donaldson took me on a tour of his ranch in the Avra Valley west of Tucson, and showed me his Brahma cattle with pride. *"There's* your answer to the cattleman's coyote problem in the Southwest," he said, "and to many of his other problems too: the Brahma. I think the Hereford should be outlawed in this part of the country." Does Donaldson have a point? I do not know. How about it, cattlemen?

Is the answer with sheep to remove them from the range, as many environmentalists propose, and raise them like poultry in totally protected enclosures as in the Y. O. Ranch experiment? What then would happen to the vast public lands of the West, once the sheep were removed? Would they be allowed to lie fallow, to renew themselves after a century of abuse? Could we stand to watch all this acreage remain unused, and hence "wasted" if we accept the Judeo-Christian attitude that land not subjugated by man is a "waste"? Would more be lost than a certain traditional flavor of the West?

One who feared so was Laney Hicks, Northern Plains representative of the Sierra Club, when I talked with her more than a year ago in Dubois, Wyoming. "I think we must help the ranchers with their predator problem," she told me, "even with poisons, if necessary. Because if ranchers give up the range, we'll lose our strongest allies against the strip-mining of the West."

Poisoned coyotes or strip mines—and then no wildlife at all! I have seen the obscenity being perpetrated at Black Mesa. My God, I sometimes think, are we forever condemned to a choice in favor of the lesser of two evils?

"The only bright spot on the horizon is the amount of coyote research that's being done," George Merrill said. Perhaps he is right. Perhaps the studies now under way will provide information which, if we will use it, will enable us to live at peace with the coyote. Much research is being done on how we can adapt the coyote into an animal who will be less destructive to man's assets.

But I do not think it fair to require that only the *coyote* change his ways. We must be ready to change ours also.

As you, the reader, will no doubt have gathered by now, this book is concerned with more than the coyote and how we might reduce coyote depredations on sheep. I certainly do not mean to denigrate in any way the importance of practical steps to de-escalate the man-coyote conflict, any more than I would try to denigrate the importance of recycling cans and paper or of building automobiles that are less wasteful of gasoline. But as I see it all such steps, however essential they may be in themselves, will remain mere palliatives unless and until we can change our fundamental thinking on man's role, rights, and responsibilities on this planet, and can alter our goals and manner of living accordingly.

This is an immense challenge to all of us, which I am not at all sure that we as a species will be ready to accept until nature rams it down our throats. And even *if* we can achieve the very broad, fundamental change in attitudes which I see as vital, it still will require the creative imagination and dedicated labor of countless men and women of vastly greater expertise than I possess, to translate the attitudinal change into a workable system of living in which we will be the stewards, not the plunderers, of the earth. The enormity of the task, and the unwillingness and incapacity of the vast majority of mankind—leaders and followers alike—to even begin to recognize it, often lead me to a state of existential despair. There sometimes seems to be so little that you or I as individuals—even well-meaning individuals—can do when there is so much to be done. And the time is getting short!

I think it was Bob Roughton who said to me one day, "I wish all our wildlife were doing as well as the coyote! There are a lot of other animals I am more worried about than him."

I agree. And one of those other animals I worry about most is Man.

We talk about the necessity to control coyote populations. But we accept the doubling every twenty or thirty years of the population of mankind. We "control" prairie dog towns. But we do nothing to halt the hideous sprawl of Denver or Phoenix, Tucson or Albuquerque. We condemn predators for taking a share of a valuable resource—our livestock. Yet we persist in wasting the earth's non-renewable resources at an ever-accelerating rate. We damn the coyote as a merciless killer. Yet we demonstrate an increasing inability to cope with the violence in ourselves, and we cannot even agree on the need for sane hand-gun-control laws.

It was Loren Eiseley who wrote, in *The Immense Journey*, ". . . The need is not really for more brains, the need is now for a gentler, a more tolerant people than those who won for us against the ice, the tiger, and the bear. The hand that hefted the ax, out of some old blind allegiance to the past fondles the machine gun as lovingly. It is a habit man will have to break to survive, but the roots go very deep."

Sometimes, in those dialogues with darkness I hold in the dead of night, I allow myself to wonder if, for the future well-being of our planet, we ought not to welcome the apocalypse recently predicted by Nobel scientist Linus Pauling, who foresaw that within the next half century an epic calamity would wipe out a good part of the human race. Unless the agent of doomsday were the hydrogen bomb, our species itself would survive—presumably somewhat chastened, somewhat humbled, determined perhaps not to repeat past mistakes.

But how many other forms of life must perish before the Armageddon?

The coyote? Perhaps not. Perhaps he will trot along his merry way, his song echoing from the mesas in the stillness of the night, "as long as the rivers shall run and the grass shall grow," as the Navajos say. Perhaps the sheepmen are right, and the coyote *will* outlast us all. And I sometimes feel that he may deserve to. Certainly we *need* the coyote, and I for one rejoice in his apparent invincibility. We need him in the same way that we need the tides, and the seasons and, yes, even such awesome manifestations of Nature's untamed power as earthquakes and hurricanes and volcanic eruptions. We need all of these as reminders that Man is not God!

But what of the coyote's cousin the wolf? A large proportion of the wolf population in Alaska's Tanana Flats area was shot to death early in 1976 by professional gunners so that more moose would become available for human hunters. A combination of hard winters, wolf predation, and heavy human hunting had seriously depleted the moose herd. And so of course the wolves had to go so that humans could have "their" moose.

Whose moose? Whose earth? Have other creatures no rights? Is the planet really ours exclusively to enjoy, manipulate, and plunder as we please? If so by what right—Divine Right, or right of might? Have we no sense of sharing, of stewardship, of kinship with other living things who are our unlucky fellow-travelers aboard Spaceship Earth? It seems not.

As George L. Small observes in his book, *The Blue Whale*,

> The tragedy of the blue whale is the reflection of an even greater one, that of man himself. What is the nature of a species that knowingly and without good reason exterminates another? How long will man persist in the belief that he is the master of this Earth rather than one of its guests? When will he learn that he is but one form of life among countless thousands, each one of which is in some way related to and dependent on all others? How long can he survive if he does not?

Some two and a half centuries have passed since Alexander Pope wrote his *Essay on Man*. In the attitude he expressed he was well ahead of his time. He was even ahead, I am afraid, of *ours*:

> Has God, thou fool! work'd solely for thy good,
> Thy joy, thy pastime, thy attire, thy food?
> Who for thy table feeds the wanton fawn,
> For him as kindly spreads the flowery lawn.
> Is it for thee the lark ascends and sings?
> Joy tunes his voice, joy elevates his wings.
> Is it for thee the linnet pours his throat?
> Loves of his own and raptures swell the note.
>
> The bounding steed you pompously bestride
> Shares with his lord the pleasure and the pride.
> Is thine alone the seed that strews the plain?
> The birds of Heav'n shall vindicate their grain.
> Thine the full harvest of the golden year?
> Part pays, and justly, the deserving steer.
> The hog that ploughs not, nor obeys thy call,
> Lives on the labours of this lord of all.
>
> Know Nature's children all divide her care;
> The fur that warms a monarch warm'd a bear.
> While Man exclaims, 'See all things for my use!'
> 'See man for mine!' replies a pamper'd goose:
> And just as short of Reason he must fall,
> Who thinks all made for one, not one for all.

Works Cited

[The numbers in brackets refer to the Chapter(s) in which the source is quoted.]

Atzert, Stephen P. *A Review of Sodium Monofluoroacetate (Compound 1080)—Its Properties, Toxicology, and Use in Predator and Rodent Control*. Washington, D.C.: U.S. Department of the Interior, February, 1971. [5]

Buys, Christian J. "Predator Control and Ranchers' Attitudes." *Environment and Behavior*, Vol. 7 No. 1, March 1975. [8]

Cain, Stanley A., John A. Kadlec, Durward L. Allen, Richard A. Cooley, Maurice G. Hornocker, A. Starker Leopold, Frederick H. Wagner. *Predator Control–1971. Report to the Council on Environmental Quality and the Department of the Interior by the Advisory Committee on Predator Control*. Ann Arbor: University of Michigan Press, 1972. [5]

Carrighar, Sally. *Wild Heritage*. Boston: Houghton Mifflin Company, 1965. [2]

Connolly, Guy E., and William M. Longhurst. *The Effects of Control on Coyote Populations: a Simulation Model*. University of California Division of Agricultural Sciences Bulletin 1872, August, 1975. [7]

Dobie, J. Frank. *The Voice of the Coyote*. Boston: Little, Brown and Company, 1949. [1, 3, 4, 5]

Eiseley, Loren. *All the Strange Hours*. New York: Charles Scribner's Sons, 1975. [2]

Eiseley, Loren. *The Firmament of Time*. New York: Atheneum, 1960. [11]

Eiseley, Loren. *The Immense Journey*. New York: Random House, 1957. [11]

Hernández, Francisco. *Nova Plantarum, Animalium et Mineralium Mexicanorum Historia*. Rome, 1651. Quoted in Dobie, *op. cit.* [5]

Horgan, Paul. *Great River: the Rio Grande in North American History*. New York: Holt, Rinehart and Winston, 1965. [9]

Krutch, Joseph Wood. *Grand Canyon: Today and All Its Yesterdays*. New York: William Sloane Associates, 1958. [2]

Laycock, George. "Travels and Travails of the Song-Dog." *Audubon*, Vol. 76, No. 5, September, 1974. [5, 9]

Leopold, A. Starker, Stanley A. Cain, Clarence M. Cottam, Ira N. Gabrielson, Thomas L. Kimball. "Predator and Rodent Control in the United States." *Transactions of North American Wildlife Conference*, 29. 1963. [5]

Mathews, John Joseph. *Wah'Kon-Tah*. Norman: University of Oklahoma Press, 1932, reprinted, 1968. [5]

McLuhan, T. C. (compiled by). *Touch the Earth: a Self-Portrait of Indian Existence*. New York: E. P. Dutton & Co., 1971. [4]

McNulty, Faith. *Must They Die? The Strange Case of the Prairie Dog and the Black-Footed Ferret*. Garden City, N.Y.: Doubleday & Company, Inc., 1970. [5]

Olsen, Gary R. and Christian J. Buys. *Perspectives on Land Use: Problems of Wildlife Management Viewed from Varying Cultural and Socio-Economic Perspectives*. Final Report to New Mexico Humanities Council, Grant #1. [8]

Roessel, Robert A. Jr., and Dillon Platero, eds. *Coyote Stories of the Navajo People*. Phoenix: Navajo Curriculum Center Press, 1968 [4]

Russell, Frank. *The Pima Indians*. Re-edition with Introduction, References, and Notes by Bernard L. Fontana. Tucson: The University of Arizona Press, 1975. [4]

Ryden, Hope. *God's Dog*. New York: Coward, McCann & Geoghegan, Inc., 1975. [3, 4, 5]

Schaller, George B. *Golden Shadows, Flying Hooves*. New York: Alfred A. Knopf, 1973. [2]

Simpson, Ruth DeEtte. "The Coyote in Southwestern Indian Tradition." *The Masterkey*, the publication of the Southwest Museum, Los Angeles, Vol. 32, No. 2, March-April 1958. [4]

Small, George L. *The Blue Whale*. New York: Columbia University Press, 1971. [11]

Twain, Mark. *Roughing It*. Edited by Rodman W. Paul. New York: Holt, Rinehart & Winston, 1953. [5]

Van Wormer, Joe. *The World of the Coyote*. Philadelphia, New York and Toronto: J. B. Lippincott Company, 1964. [3]

Appendix A

PROGRESS IN COYOTE DEPREDATIONS RESEARCH*
by
SAMUEL B. LINHART

I appreciate the opportunity to discuss the status of coyote depredations control research at this meeting and would first like to discuss the history of research in this area before reviewing the status of ongoing work.

Coyote depredations research from the end of World War II to about 1960 was characterized by a low level of funding and manpower, and by rather primitive research facilities. The primary emphasis was on the development of predacides—without the benefit of input from allied specialty areas such as pharmacology and physiology. Legislative restrictions or guidelines were generally lacking and relatively little effort was directed toward the effects upon non-target species or the environment. This period was further characterized by a lack of public interest in the coyote-sheep problem and few individuals felt the need to document and record the results of routine predator control activities. This lack of documentation became painfully evident last spring in the process of putting together an application for the registration of the M-44 device. To mention but a single example, to date we still lack any quantitative data as to the effects of 1080 bait stations on damage suppression, on coyote populations, on non-target species, or the extent, if any, of environmental hazard associated with their use. Progress following World War II, at least on predacide development, was fairly rapid since the above factors were largely ignored and regulatory procedures were generally lacking.

The creation of the Leopold Committee in 1964 and the Cain Committee in 1971 was evidence of a growing public concern and awareness of predator control activities in western United States. These committees, their recommendations, and the resultant furor, were merely one indication of a growing public interest in the potential hazards related to the application of pesticides. The little old ladies in tennis shoes started changing into little old ladies in combat boots and the terms "ecology" and "environment" came into vogue.

*This paper by Dr. Linhart of the Fish and Wildlife Service's Denver Wildlife Research Center was presented at the Great Plains Wildlife Damage Control Workshop, Manhattan, Kansas, in December 1975.

Concurrent with increased public interest in the coyote-livestock problem in the 1960's was a major change in emphasis of depredations control research undertaken by the Fish and Wildlife Service. Whereas emphasis prior to the 1960's had been on the development of toxicants, there was a redirection of our research efforts into the area of antifertility agents to reduce coyote population in areas where depredations were a problem. This research, almost to the exclusion of all other areas of inquiry, was conducted from about 1960 to 1967, during which time intensive field studies were undertaken in a number of the western states. This work consisted of the aerial or ground application of tallow drop baits containing stilbestrol, a synthetic estrogen, with recovery of coyotes from study areas to determine their effects upon productivity. In general, the results of these tests were negative and suggested that (1) an inadequate proportion of the coyote populations were exposed to and/or consumed baits, (2) that drop baits disappeared in a matter of days because of the scavenging efficiency of rodents and birds, (3) that better chemosterilants effective over a larger segment of the breeding season were needed, (4) that better bait attractants were necessary, and (5) that the effort involved to determine effects on non-target species for registration purposes would be considerable. Although the concept of reproductive suppression to bring about population reduction remains valid to this day, the methodology available at that time did not appear adequate to achieve our objectives. However, one positive aspect of this research was the generation of ancillary, but necessary, studies such as the development of markers or tracers to determine which individual animals consumed drop baits, extensive laboratory work on the basic reproductive biology of the coyote, initiation of field studies to determine indices of relative abundance, fabrication of telemetric equipment to determine coyote movement, and the collection of data which would subsequently be used to characterize the population dynamics of this species. Such data, we now know, are essential to understanding the effects of control upon coyote populations.

Although it was generally assumed at the time these studies were conducted that general population suppression was the primary means by which depredations could be reduced, increasing thought was being given to the alternate approach of controlling only individual coyotes causing damage to a specific sheep flock or band. The question of general or preventive population suppression versus control of individual problem coyotes still requires investigation.

Public concern, whether legitimate or not, resulted in the issuance of an Executive Order in February of 1972 that prohibited the use of toxicants on federal lands or by federal employees. This Order rapidly and dramatically changed the entire coyote management picture as well as supportive research, and emphasis was redirected toward development of nonlethal means of suppressing depredations. The restrictions imposed by the 1972 Executive Order generated additional controversy between the anti- and pro-coyote factions, and this, in turn, resulted in increased funding to search for alternate methods of reducing depredations.

Up until the 1972 Executive Order, there was little university-sponsored research on the problem; nevertheless, the sudden availability of funds for contract work resulted in university involvement. The Agricultural Research Service, U.S. Dept. of Agriculture, received funds following the Executive Order to work on the problem, and the Fish and Wildlife Service allotment for research in this area was also increased. The ARS funds, restricted until very recently to research on nonlethal methods, was initially channelled into four major contract areas. These included a contract to the University of Wyoming to look at gustatory repellents, a second to Colorado State University to look at olfactory repellents, a third to South Dakota State University to research the potential of repellent sounds, and a fourth to Texas A&M at San Angelo to explore several possibilities, including repellents and fencing. At present, in-house work, as well as contractual studies are being funded by the ARS.

The U.S. Fish and Wildlife Service, at our Denver Research Center, increased the Section of Predator Damage staff, and began looking for alternate means of reducing depredations, both in-house and via contractual studies. The number of university and state-funded investigations also increased at this time.

The proliferation of coyote studies generated by increased public interest and new funding made it increasingly difficult for researchers to keep abreast of "who was doing what," and the problem of coordination and information exchange became acute. In response to this situation, the DWRC began distribution of a Coyote Research Newsletter in June of 1973. The initial issue listed a total of 92 separate coyote research projects grouped into three major categories—biology, ecology, and behavior; damage assessment; and depredations control. A subsequent issue included an additional 35 research projects. The proliferation of studies related to the coyote-sheep problem increased further by a congressional appropriation to the Economic Research Service, USDA, which was directed to undertake a major study to determine the magnitude of the depredations problem within the western sheep industry, to measure major factors influencing it, and to attempt to model coyote populations and the effects of control upon them. Some indication as to interest in coyote-sheep investigations is given by the current mailing list for the Coyote Research Newsletter which now approaches 900.

I would like to take my remaining time to discuss some major areas of ongoing research and to review "who is doing what." Several attempts are now being made to coordinate research on the coyote-sheep problem. For example, the USDA's Technical Committee of Western Regional Research Project W-123 deals with the coyote-sheep problem and meets annually to review progress and to discuss mutual interests. The last meeting was held in Uvalde, Texas, at the Texas A&M University Research and Extension Center. These meetings serve a very useful function and should be continued. Secondly, the Coyote Research Newsletter will continue to provide current information to researchers. Third, I hope that sometime in the near future a second Coyote Research Workshop, such as that sponsored by the Fish and Wildlife Service in Denver in November of 1974, will be scheduled so that researchers can exchange information. Approximately 75 papers were presented and 250-300 people attended the last Workshop. Fourth, the American Society for Testing and Materials is developing standards or recommended procedures for the testing of biocides. One deals with the evaluation of predacides, a second with sodium cyanide, a third with the use of 1080, and a fourth is being considered that deals with the use of chemosterilants for suppression of carnivore populations. Such standards, providing they do not discourage innovative concepts or approaches, should be helpful.

Keeping up to date on the rapidly expanding body of coyote literature is a real problem. In this respect, we should see several very helpful publications come out in the near future. For example, the University of Colorado has assembled a mimeographed coyote bibliography and is now in the process of editing a book dealing specifically with the coyote. This book will be published by Academic Press. The Denver Center currently has an ADP search and retrieval system containing some 4000 citations on predator-related subjects. Each citation is indexed by one or more of 93 descriptor words. This bibliography, which is periodically updated, is tied into the Proprietory Computer Systems computer in California and can be searched by subject. While the Denver Center presently has no means of accepting payment for searches, arrangements are being made for investigators outside the Fish and Wildlife Service to obtain searches from the Denver Public Library in Denver, Colorado. Such searches will be available for a nominal fee. The Fish and Wildlife Reference Service, also in Denver, is an excellent source of data and information generated by state Pittman-Robertson and Dingell-Johnson-supported research.

The Denver Center has made arrangements for Carol Snow, who authored a number of wildlife management publications for the Bureau of Land Management, to write an extensive literature review of the coyote. In addition, one of our biologists, Frank Turkowski, is in the process of compiling a 3000 citation bibliography on the wild canids of North America. We hope to have this bibliography, as well as Snow's review, published as a Fish and Wildlife Service Special Scientific Report. All the efforts mentioned above should facilitate literature review and better coordination and exchange of information.

I would now like to review current research on nonlethal techniques for reduction of coyote depredations upon sheep. Electric fencing has been used in Kansas and Illinois to exclude coyotes and according to some individuals is quite effective. Maurice Shelton, with Texas A&M University in San Angelo, Texas, has made a concerted effort to reduce coyote predation by means of fencing and appears to have had little success. The Denver Center has funded a contract to Oregon State University to evaluate different configurations of net-wire and/or electric fences. The results of this study should be available by the spring of 1976. I understand that the ARS research facility in Clay Center, Nebraska, is also exploring the use of fencing to protect sheep. Fencing, whether it be simply mechanical exclusion or electrical, may offer potential under certain conditions, but additional field studies using accurate damage assessment techniques are needed.

To my knowledge, research on the use of chemosterilants to suppress coyote populations is presently being conducted by a single agency, the ARS experimental station in DuBois, Idaho. The station has constructed coyote holding facilities and is exploring the effects of several compounds on spermatogenesis in the male coyote. While only one investigator is actively engaged in this work, several recent modeling studies indicate that the concept of suppressing populations by chemosterilants is a valid one, despite the problems encountered by the USFWS when we attempted to demonstrate field efficacy several years ago. It is to be hoped that advanced technology and field methodology will produce better results.

Work on coyote repellents has been conducted by a number of individuals. Both the Denver Center and the University of California (Davis) have devised standardized techniques for initial screening of candidate repellents using penned coyotes. To date, the Denver Center has only field tested a single compound, Cinnamic Aldehyde, via a contract with North Dakota State University. The results of this investigation were inconclusive. An ARS-funded contract with the University of Wyoming is concerned with a number of candidate repellents and several hundred compounds have been evaluated. Two compounds, Bitrex and Norcapsaicin, have been tested using confined sheep and coyotes. Norcapsaicin offers more positive protection than does Bitrex. Researchers with ARS at Albany, California and DuBois, Idaho, and at Texas A&M University at San Angelo, Texas are collaborating to test several candidate repellents. These include bone oil, Bitrex, red pepper extract, and an extremely bitter plant extract, plictran. Both pen and field evaluations are continuing. Despite efforts at Colorado State University and elsewhere, none of the odor-based repellents have shown as much potential as have those based on gustatory response. From results thus far, it appears we have a way to go before a functional coyote repellent, if such exists, is developed.

Limited effort has been directed toward the use of aversive agents to deter coyotes from attacking sheep. While the principle of aversive conditioning is well documented, only a few studies related to the coyote-sheep problem have been conducted. Carl Gustavson, currently at Eastern Washington State University at Cheney, has reported the successful use of lithium chloride baits to avert coyotes from attacking sheep. Although there seems to be some disagreement as to the methodology employed and the conclusions drawn from this work, it is nonetheless encouraging to note that research is underway in this area. In this regard, the Denver Center has funded a two-year contract to Colorado State University to develop a suitable protocol for evaluating aversive agents and to test several compounds that may offer potential. This contract was recently initiated and no results are available as yet.

There has been some confusion as to what constitutes a repellent versus an aversive agent. We consider a compound an aversive agent when it induces physiological illness in a coyote and after an initial exposure or two causes a learned avoidance of an associated food or prey species; whereas, repellents require no initial learning but sufficiently disturb or irritate specific sensory systems as to repel the coyote away from treated prey. In addition to the CSU contract, we hope to conduct laboratory investigations at Denver to correlate dosing regimes with onset of illness, and induction of aversion effects.

Several recently published papers have shown that fatty acids are a major chemical component in the anal gland secretions of the dog, monkey, and coyote. Another recent paper dealt with the presence of these same fatty acids in human vaginal secretions. To what extent these acids serve a pheromone function remains to be demonstrated; however, interestingly enough several are used in scent formulations by trappers. This information and the results of limited field tests have proved extremely interesting to our researchers in Denver and have prompted a study aimed at identifying fractions of substances eliciting specific coyote behaviors. Biochemical analyses, behavioral responses of captive coyotes, and field evaluation should result in development of better coyote attractants. Concurrently, researchers with the ARS at Albany, California, and in cooperation with the University of California, are identifying fractions of coyote urine and attempting to learn which elicit identifiable responses.

Over a period of many decades, trappers have developed some very effective attractants. We hope to take advantage of their knowledge as well as modern and systematic techniques in the development of superior attractants.

Another area of inquiry is that underway at South Dakota State University where the auditory sensitivity of the coyote, the appropriate frequencies for sound-repelling devices, and the extent to which these devices inhibit coyotes from killing livestock are being looked at. Electric shocking collars to reduce coyote predation have also been investigated at SDS. Letters of inquiry regarding sound as deterrents have also been received from several people in the state of Washington who want to do work in this area. The DWRC will shortly initiate work on frightening devices—including the use of lights and sound. To date, little effort has been aimed at this particular approach.

Nearly all the nonlethal techniques for reducing coyote depredations on sheep discussed thus far involve attempts to elicit certain behavioral responses from the coyote—and these responses are dependent upon olfactory, auditory or visual interactions. Yet a review of the literature reveals that very little data have been collected as to the sensory parameters of the coyote and their relative importance in eliciting or inhibiting predatory response. Because of their obvious importance to the development of nonlethal control methods, the Denver Center has contracted with Colorado State University to look at their relative importance in eliciting and inhibiting predatory attack.

The Denver Center has also funded a contract with the University of Wisconsin to study hypersensitivity in canids and to assess its potential for coyote control. In theory, this approach involves an attempt to make coyotes "allergic" to ovine antigens so that a depredating coyote would react in the form of a fatal anaphylactic shock or a sufficiently painful response as to constitute an aversive stimulus. This study has not yet proceeded to the point where its possible application can be assessed.

A final research project which may or may not fall into the nonlethal category involves the use of guard dogs to protect sheep. There are at least three breeds of large dogs that have been used in Europe and Asia for hundreds of years to guard sheep and goat flocks from attacks by wolves. One of these, the Komondor, has been selected by the Denver Center for a low-profile, long-term study. By low-profile, I mean that a relatively small percent of our time and funds will be directed towards this approach. We are currently in the process of gathering all available information on these dogs, including talking to individuals who have used them under field conditions; we will shortly acquire several dogs and formulate a specific research protocol.

In the past two years, we have seen the experimental use of the lethal

M-44 device (using sodium cyanide), its subsequent registration by the Environmental Protection Agency, and modification of the 1972 Executive Order permitting the use of sodium cyanide as a toxicant for predator control. These events seem to indicate a swing of the pendulum toward the renewed use of toxicants. However, their use will be much more tightly controlled and more closely scrutinized by the public and environmental groups than was the case 5, 10, or 20 years ago. In this regard, aside from the M-44 studies by the FWS and states issued Experimental Use Permits by the EPA, little recent effort has gone into the use of toxicants.

As most of you know, a year ago the Denver Center initiated a study to evaluate the toxic sheep collar, a toxicant-filled bladder or series of packets placed on sheep being attacked by coyotes. This approach eliminates only depredating animals and has, in general, been well received by the public. Such a collar was patented several years ago by Roy McBride of Alpine, Texas who reported excellent results using a rubber bladder filled with a solution of compound 1080. We elected to test a sodium cyanide formulation in a collar fabricated of several PVC packets. Following successful pen tests at Denver, field studies were conducted in North Dakota, Montana, and Texas last fall and we can now state unequivocally that while the "operation" was a success, all the "patients" (i.e. coyotes) survived! We found that sodium cyanide because of its odor and/or taste apparently resulted in sublethal aversion upon initial strike and that an improved collar design was needed. We also attempted to tether all collared lambs and believe that in many instances tethering alone deterred coyotes from attacking. We are now in the process of looking at alternate toxicants and collar designs and hope to resume field tests this winter or next spring.

In addition to the toxic sheep collar, there may be other means of selectively applying toxicants to reduce depredations. For example, the search for alternate toxicants to use in the collar may have some spin-off benefits. A study to be initiated next spring may answer some of the questions as to efficacy and selectivity of baits and baiting. DWRC investigators using baits containing physiological markers, will be investigating the probable effects of toxic or chemosterilant baiting procedures on target and non-target species. We hope these studies will provide us with *factual* information that has been sadly lacking and that will perhaps be of value in making predator management decisions in the future.

As you can see, numerous investigators are looking at a wide variety of approaches to the problem. Such diversity is indeed a wise course of action since experience has shown that no single approach, or any one method, is adequate for all conditions. Predation upon livestock occurs under a multiplicity of circumstances involving many different situations with respect to prey species, habitat, vegetative conditions, weather and climatic variations—not to mention human whims and influences. It is therefore essential that research advances on as many fronts as possible. New techniques, acceptable to the livestock producer and public alike, must be supported by carefully collected data showing that such procedures are effective and are safe for humans, non-target species, and the environment.

While I have attempted to review ongoing depredations control research, I would be remiss if I failed to briefly mention the many damage assessment, ecology, and behavioral studies underway. For example, the damage assessment studies of the Economic Research Service group, of Wagner, Bowns, Nelson and Curley in Utah, Early in Idaho, Howard and DeLorenzo in New Mexico, Henne in Montana, Klebenow and MacAdoo in Nevada, Nesse in California, Beasom in Texas and those by Tigner and Nass with the Denver Center, have contributed greatly to our knowledge of sheep losses under varied conditions, whether from disease, poisonous plants, mismothering, still births, or predators. These data have significantly modified, I believe, the attitude of many of the environmental groups who now concede that under certain conditions and in certain locations coyotes do cause significant losses to the sheep producer.

Study of the ecology of the coyote is rapidly increasing our data base and this better understanding of movement, home range, population dynamics, natality, and causes of mortality is essential to the development of depredations control methods. As an example, three papers, one by Sheriff and Cringan (Colorado State University), a second by Pyle (University of Washington), and a third by Connolly (Denver Wildlife Research Center) have dealt with coyote population models and response to various levels and types of control. This information should prove extremely helpful to researchers and managers.

Several papers have also dealt with methods of determining relative coyote abundance, including the use of siren-induced howl counts, aerial census, and response to scent stations. While admittedly these techniques can be improved upon, we are, at least, beginning to get a handle on year-to-year variations in relative coyote densities and their response to control measures—knowledge crucial to the management of any wildlife species.

Finally, I must confess that this review has been somewhat frustrating. For despite all efforts in the past several years, we have yet to see a really new technique reach the point where it is of benefit to the livestock producer. Perhaps this apparent lack of progress can be attributed to two factors: First, the renewed interest in depredations control research really came about only two or three years ago and such studies take time. And second, the coyote is an extremely intelligent and adaptable animal and seems to thrive despite our best efforts. Let's hope that one or more of the approaches I've mentioned will bear fruit. One thing seems apparent. Over a relatively short period of time, the philosophy and attitudes of the public have changed and future efforts to manage coyotes must be justified on the basis of demonstrated losses, the use of selective control methods, and knowledge as to their potential hazard to man, non-target species, and the environment.

Appendix B

Table 1
Coyotes Killed by ADC

Year	Arizona	New Mexico
1964	4,119	5,940
1965	3,257	4,634
1966	1,963	2,804
1967	1,359	3,402
1968	1,973	3,668
1969	1,864	3,521
1970	2,085	4,352
1971	1,494	4,098
1972	1,080	3,529
1973	963	3,590
1974	997	5,465
1975	1,141*	5,200

SOURCE: Animal Damage Control

* Plus 177 coyotes taken on the Navajo Reservation with the M-44 cyanide gun. In 1975 President Ford relaxed the executive order on toxicants to allow operational use of the new type of cyanide gun, and the Environmental Protection Agency re-registered sodium cyanide for use in predator control.

Table 2
Arizona
Animals Taken and Method Used, F.Y. 1975

Animals	Trapped	Shot	Snared	Total
Black Bear	7	1	1	9
Bobcat	46	1		47
Coyote	1,020	117	4	1,141
Mt. Lion	4			4
Beaver	2			2
Badger	29			29
Fox	93			93
Raccoon	5			5
Skunk	84			84
Total	1,290	119	5	1,414

SOURCE: Animal Damage Control
NOTE: This table does not include animals taken from the Navajo Reservation by employees of the Navajo Tribe. This information is shown in the following table.

Table 3
Arizona
Navajo Predatory Animal Control
Animals Taken and Method Used, F.Y. 1975

Animals	Trapped	Shot	M-44ed	Denned	Total
Bobcat	123	25		2	150
Coyote	330	26	177	53	508
Fox	200	10	47	4	227
Dogs	644	398	44		1,071
Badger	37	2			39
Porcupine	46	1	1		48
Raccoon	2				2
Skunk	40	6	4		46
Total	1,422	468	273	59	2,091

SOURCE: Animal Damage Control

Table 4
New Mexico
Animals Taken and Methods Used, F.Y. 1975

Species*	Traps	Cyanide Ejector**	Guns	Snare	Dens	Air-plane	Dogs	Total
Badger	221	--	1	10	--	--	--	232
Bobcat	245	--	3	18	--	7	3	276
Coyotes	3,063	180	237	357	183	1,179	1	5,200
Fox	383	29	3	7	--	--	--	422
Porcupine	610	--	12	40	--	--	--	662
Raccoon	56	1	--	--	--	--	--	57
Skunk	446	15	2	--	--	--	--	463
Total	5,024	225	258	432	183	1,186	4	7,312

SOURCE: Animal Damage Control
*In addition to the above trapped, the following were released: 7 bobcats, 25 dogs, 1 house cat, 1 ringtail, 1 deer, 24 fox, 41 badgers.

**In addition to the above by cyanide ejector, the following were taken: 1 crow, 1 feral dog, 1 calf.

Table 5
Arizona Cattle Growers' Association
Predator Losses Survey Recap
from 11/1/74 to 11/1/75

	Number of Deaths Reported	Average Price per Head	Dollar Amount of Death Losses
Calves	2,476	$106.86	$264,585.36
Bulls	--	--	--
Cows	64	$182.54	11,682.56
Yearlings	142	$199.45	28,321.90
Horses	1	$300.00	300.00
Colts	13	$300.00	3,900.00
Sheep	2,003	$ 60.00	120,180.00
Poultry	445.5	$ 1.50	668.25
Pig	1	$100.00	100.00
Total Dollar Amount of Losses			$429,738.07

NOTE: 297 members responded with predator losses (459 individual cases of depredations reported); 215 members responded with no losses from predators; 404 members who own cattle did not respond to survey.

Breakdown of Deaths Reported by Various Predators

	Calves	Bulls	Cows	Yearlings	Horses	Colts	Sheep	Poultry	Pigs
Coyotes	1220.5	--	9	3	--	--	1544	377.5	1
Bobcats	88.5	--	--	--	--	--	--	61	--
Lions	776.5	--	28	100	1	13	7	--	--
Bears	203.5	--	19	30	--	--	--	--	--
Wild dogs	159	--	8	5	--	--	102	7	--
Dom. dogs	28	--	--	4	--	--	350	--	--
Totals	2476.0	--	64	142	1	13	2003	445.5	1

SOURCE: Arizona Cattle Growers' Association